BEYOND THE RAINY SEASON
DISCOVERING LONAVALA:

UNVEILING ECOLOGICAL, PHYSICAL AND ECONOMICAL ASPECTS.

Prof. Suchetha Mathew Thazhathil,
Prof. Viji Nair, Prof. Swechcha Gosavi

notionpress
.com

INDIA • SINGAPORE • MALAYSIA

ISBN
Paperback 979-8-89744-428-1
Hardcase 979-8-89744-429-8

With deep admiration and respect, this book is dedicated to architects, educators, researchers and students who shape our world with vision and creativity. Your passion for design, devotion to sustainability and pursuit of excellence illuminate spaces, inspire minds and transform the built environment for generations to come.

MENTORS & EDITORS

Prof. Suchetha Mathew Thazhathil, former Professor and Principal of Vishwaniketan College of Architecture, Arts & Design, is an environmental architect with over 25 years of professional experience. With more than 15 years of teaching in Bachelor's and Master's programs at leading architectural institutions in Mumbai, she has made significant contributions to academia. Currently pursuing a Ph.D. at the National Institute of Technology (NIT) Calicut, her research explores the impact of the built environment on mental health, focusing on how design influences human wellbeing.

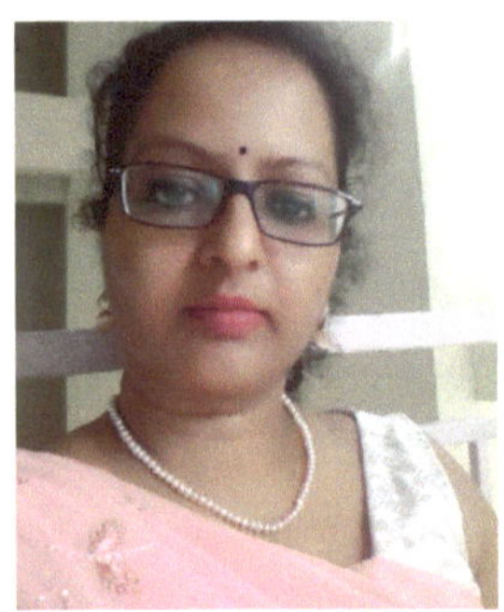

Prof. Viji Nair formerly working as Associate professor at Vishwaniketan college of Architecture Arts & Design, holds a Bachelor's in Architecture and a Master's in Architecture with a specialization in Project Management. She is an Indian Green Building Council Accredited Professional with 14 years of experience in diverse projects, including interiors and residential townships. She has more than 8 years of teaching experience with keen interests in sustainable practices and urban studies.

Prof. Swechcha Gosavi is an architect with a Bachelor's in Architecture and Master's in Architecture with a specialization in Urban Design, and over 10 years of experience across diverse architectural and interior design projects. She currently serves as an Assistant Professor at Vishwaniketan College of Architecture Arts & Design, where she has been teaching for more than 3 years. Her professional interests include sustainable urban development, the integration of environmental considerations into design, and the creation of socially responsive urban spaces.

ACKNOWLEDGEMENT

We extend our deepest gratitude to Vishwaniketan College of Architecture, Art & Design (VCAAD) for their unwavering support and encouragement throughout the creation of this book. The resources, expertise, and collaborative spirit provided by the institution were invaluable to this project. We are especially grateful for the support and insights from our colleagues, who generously shared their time and knowledge, enriching the content and ensuring its quality. Without the institution's commitment to fostering an environment of learning and discovery, this work would not have been possible. Thank you for being an integral part of this journey.

FOREWORD

This book explores the captivating realms of Lonavala, a major hill station in Maharashtra, India. The book focuses on the ecological aspects, physical attributes, economic significance that shape Lonavala. This comprehensive exploration takes readers on a remarkable journey through the heart of Lonavala's essence, revealing the intricate tapestry that weaves together its past, present and future.

The authors have skilfully unveiled the cities iconic structures, from ancient temples and historic monuments to modern architectural typology, illustrating the regions evolution over time. Through vivid descriptions and stunning visuals, it illustrates the architectural brilliance that graces every corner of Lonavala.

This book unravels the geological features that paint Lonavala's landscape and the ecological significance of preserving its natural beauty for generations to come. Additionally, the authors astutely shed light on the economic dynamics, showcasing how industries, markets and tourism have contributed to Lonavala's growth and prosperity.

It is a meticulous research work with thoughtful insights which has resulted in a holistic portrayal of Lonavala's multifaceted identity. This book is more than just a tribute to the city's architectural splendours and is a testament to the resilience and spirit of its people who have nurtured Lonavala into a thriving economic hub it is today. It would serve as an invaluable resource for architects, urban planners, researchers and all who seek to unravel the essence of this magnificent city.

CONTENT

Image 1: Loanvala dam

INTRODUCTION

Lonavala, set against the backdrop of the Sahyadri Mountain Range in Maharashtra, India, is a destination of breath taking beauty and charm. Situated between the bustling cities of Mumbai and Pune, it serves as a convenient and popular tourist attraction for travellers from both metropolises.

Surrounded by lush green valleys and enjoying a pleasant climate year-round, this picturesque hill station undergoes a magical transformation during the monsoon season, when the landscape is draped in vibrant greenery, and waterfalls cascade down the hills, making it a haven for nature lovers and adventure enthusiasts alike.

The name "Lonavala" is derived from the Sanskrit word lonavli, meaning "group of caves," a fitting reference to the nearby ancient cave complexes such as Karla Caves, Bhaja Caves, and Bedsa Caves, all of which lie in close proximity to the town.

Present-day Lonavala was once part of the Seuna (Yadava) Dynasty. Later, recognizing its strategic significance, the Mughals controlled the region for an extended period. The forts in the area, along with the "Mavala" warriors, played a crucial role in the history of both the Maratha Empire and the Peshwas. In 1871, the hill stations of Lonavala and Khandala were discovered by Lord Elphinstone, who served as the Governor of Bombay Presidency at that time.

1.Ecological Landscape

This section explores the intricate ecological framework of Lonavala, an area where nature's grandeur thrives within its diverse ecosystems. Understanding the delicate balance of this region's biodiversity requires a detailed approach through ecological mapping. By analyzing its land cover and evaluating the significance of its watershed, this study seeks to highlight Lonavala's environmental importance and underscore the necessity for sustainable conservation measures.

Lonavala, a renowned hill station celebrated for its natural beauty and tranquil landscapes, serves as a prime subject for ecological investigation. This chapter delves into the unique ecological elements that make Lonavala a valuable destination for nature enthusiasts, tourists, and researchers alike. The region's ecosystem

Figure 1: Location shown in political map of India

is a harmonious blend of varied land cover, complex topography, watersheds, and diverse flora and fauna, making it an ideal case study for environmental exploration.

A comprehensive understanding of Lonavala's land cover is fundamental to appreciating its ecological diversity. From dense forests and undulating hills to anthropogenic developments, the land cover serves as a diagnostic tool for assessing the region's environmental health. Through the application of advanced mapping technologies, this section uncovers valuable data on the distribution of forested areas, agricultural lands, water bodies, and urban infrastructure. By tracing Lonavala's landscape transformation over time—from untouched wilderness to a rapidly urbanizing hub—this study reveals the profound impacts of human intervention and stresses the importance of sustainable development in maintaining ecological balance.

Lonavala's topography, a complex matrix of peaks and valleys, forms the foundation of its ecological wealth. From the towering heights of the Sahyadri range to the verdant, fertile valleys below, this analysis highlights the region's topographical significance in fostering ecological diversity.

The watershed system in Lonavala plays a crucial role in preserving ecological stability. This section delves into the hydrological network of rivers, streams, and lakes that support the area's biodiversity and human population. The importance of implementing responsible watershed management strategies is emphasized, as they are essential to maintaining this delicate ecological balance.

This chapter takes an in-depth look at the species inhabiting the area, emphasizing the critical need for conservation efforts to protect these natural treasures. Through this technical examination, the ecological richness of Lonavala is laid bare, underscoring the urgent need for sustainable practices to ensure its preservation for future generations. Ecological mapping plays a crucial role in identifying ecologically sensitive zones, allowing authorities to implement protective measures such as establishing nature reserves, wildlife sanctuaries, or green belts to conserve biodiversity. Additionally, it supports responsible infrastructure development by minimizing environmental disruption and promoting eco-friendly practices, ensuring that the natural allure and ecological significance of Lonavala remain intact for future generations.

2. Urban Morphology

This chapter embarks on a comprehensive geographical exploration of Lonavala's physical attributes. Lonavala's topography plays a crucial role in shaping its unique microclimates and ecosystems. By analyzing the geological forces that sculpted this landscape, we gain deeper insights into its geological history and the impact of natural phenomena on the region's physical formation.

Lonavala's physical landscape also includes the town's urban growth, infrastructure, and architectural evolution. This chapter chronicles Lonavala's transformation from a quaint hamlet into a thriving urban hub. Its evolution from a small village to a bustling centre of activity highlights the region's economic and cultural significance. We trace the pivotal milestones in Lonavala's urban journey, examining the factors that have contributed to its rapid expansion and development. Through physical mapping, we uncover the major routes and nodes that connect Lonavala's diverse regions. These urban arteries form the core of the town's transportation and commercial networks, influencing its development and accessibility. This section also marks Lonavala, showcasing essential landmarks, transportation options, and tourist attractions.

The allure of Lonavala shifts with the changing seasons, and here we explore the optimal times to visit this hill station to experience its beauty in full bloom. Whether it is the verdant monsoon landscapes or the serene winter charm, each season offers a unique appeal to visitors.

Lonavala's architecture is a rich blend of historical influences and contemporary designs. By examining the town's architectural development, we gain valuable insights into its cultural heritage and the significance of its building topology. The fusion of tradition and modernity in Lonavala's architecture reflects its vibrant history and ongoing transformation, adding yet another layer to the town's multifaceted identity.

3. Economic Mapping

This chapter provides an analytical overview of Lonavala's economic landscape, extending beyond its well-known tourism appeal. It explores the region's economic development, traditional industries, and its adaptive strategies in response to evolving market dynamics. From agriculture and artisanal craftsmanship to emerging industries and investments, we dissect the key economic drivers sustaining Lonavala's growth. Additionally, this section assesses the balance between economic expansion and the preservation of the region's natural environment, addressing the challenges and opportunities for sustainable development.

Lonavala's economic framework is a fundamental aspect of its regional identity. This chapter evaluates the commercial activities, local employment, and industrial sectors that contribute to the town's economic dynamism. Lonavala's strategic location between major urban centers and its natural beauty have attracted a wide range of commercial ventures. We examine the economic epicenters fuelling trade, tourism, and investment, shaping the region's fiscal environment.

In addition, this section outlines key resources and amenities available to residents and visitors, such as local markets, healthcare facilities, educational institutions, and recreational areas, ensuring comprehensive accessibility and quality of life for all stakeholders.

The local population plays an essential role in defining Lonavala's economic culture. This mapping explore the livelihoods and occupational structures of its residents, offering insights into their contribution to the town's economic fabric. By examining the interplay between traditional occupations and modern economic shifts, we gain a deeper understanding of how local communities adapt to changing economic circumstances.

Lonavala's industrial landscape is also a critical component of its economic diversity. Also, analyze the primary industries in the region, evaluating their influence on employment rates, economic growth, and long-term sustainability. The industrial sector's impact on environmental and social sustainability is a key focus of this analysis

Image 2: Waterfall in Loanvala

1.Ecological Landscape

Amidst the rolling hills and verdant landscapes of Lonavala lies a hidden realm of ecological wonders waiting to be explored. As the sun rises over mist-covered valleys and sets beyond the horizon, the intricate web of life here intertwines in harmony, narrating a captivating story of nature's resilience and beauty

Lonavala, a picturesque hill station nestled between two major cities Mumbai and Pune of Maharashtra, India, is celebrated for its lush green vistas, cascading waterfalls, and temperate climate. However, with the increasing influx of tourists and rapid urbanization, the region's delicate ecological balance is facing growing threats. Ecological mapping involves the systematic assessment, documentation, and analysis of the area's ecosystems, enabling informed decision-making and promoting sustainable development.

Lonavala is part of the Western Ghats, one of the world's biodiversity hotspots. This region is home to a wide variety of endemic species of flora and fauna. The rich biodiversity includes variety of ecosystems that support species like tiger, deer, a myriad of bird species, and diverse plant life. The hills and forests of Lonavala are critical to maintaining this biodiversity, acting as a natural corridor for wildlife and a refuge for many endangered species.

Lonavala, along with its twin hill station Khandala, is situated at an elevation of 622 meters (2,014 ft) above sea level in the Sahyadri ranges, which demarcate the Deccan Plateau from the Konkan Coast. Together, these hill stations span an approximate area of 38 square kilometers (15 sq. mi). The region sees a peak in tourism during the monsoon season when its natural beauty is at its zenith.

Figure 2 : Land covered area in Lonavala

1.1 Mapping the Land Cover

Lonavala Town This is recognized as the central commercial area of the town, featuring a wide array of shops, markets, restaurants, and essential local amenities. Well-connected to various transportation options, it serves as a convenient hub for tourists seeking easy access to nearby attractions and key points of interest.

Khandala: Located next to Lonavala, is a well-known hill station that features similar topographical characteristics. Famous for its verdant landscapes, mist-covered mountains, and breathtaking viewpoints, Khandala is a popular destination for nature lovers and trekking enthusiasts alike.

Tungarli Lake And Dam: Situated near Tungarli village, Tungarli lake and dam is a picturesque destination known for its tranquil setting and scenic views of the surrounding hills. The reservoir is a popular spot for picnics and boating activities.

Rajmachi Fort: Located near Lonavala, is a renowned trekking destination, offering stunning views of the Sahyadri mountain range and its valleys. This fortified hill fort holds historical significance, drawing both adventure seekers and history enthusiasts.

Bhushi Dam: Situated near the INS Shivaji naval base, is a favored tourist attraction famous for its cascading waterfalls and peaceful ambiance. Visitors often walk along the dam's steps and enjoy the refreshing waters.

Valvan Dam& Garden: Located a few kilometers from Lonavala, is a scenic reservoir surrounded by beautifully landscaped gardens. This area is ideal for leisurely walks, allowing visitors to relax and appreciate nature's beauty.

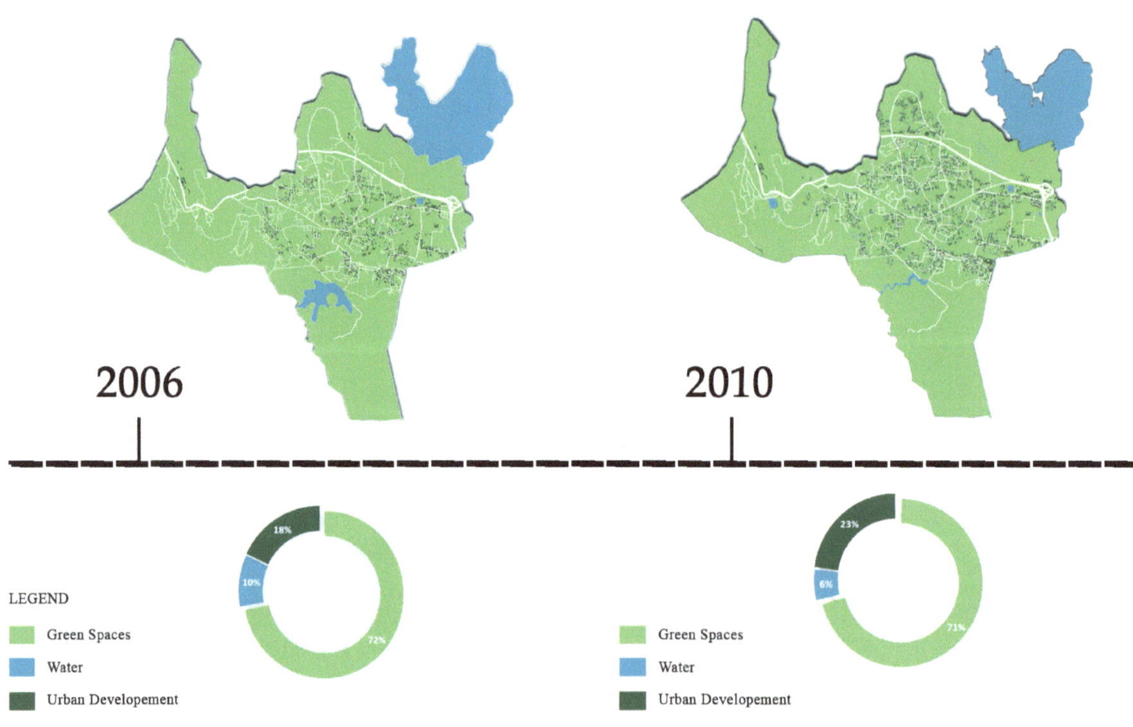

2006
2010
18%
10%
72%
23%
6%
71%
LEGEND
Green Spaces
Water
Urban Developement
Green Spaces
Water
Urban Developement

1.2 Wilderness to Urbanization: Changes in land-use across time

Lonavala has experienced significant ecological changes over the years, primarily influenced by human activities. Natural habitats and green areas have steadily given way to residential developments, commercial establishments, and expanding infrastructure. The development of roads, highways, and public transport systems has further disrupted local ecosystems, causing habitat fragmentation and altering the region's land cover patterns.

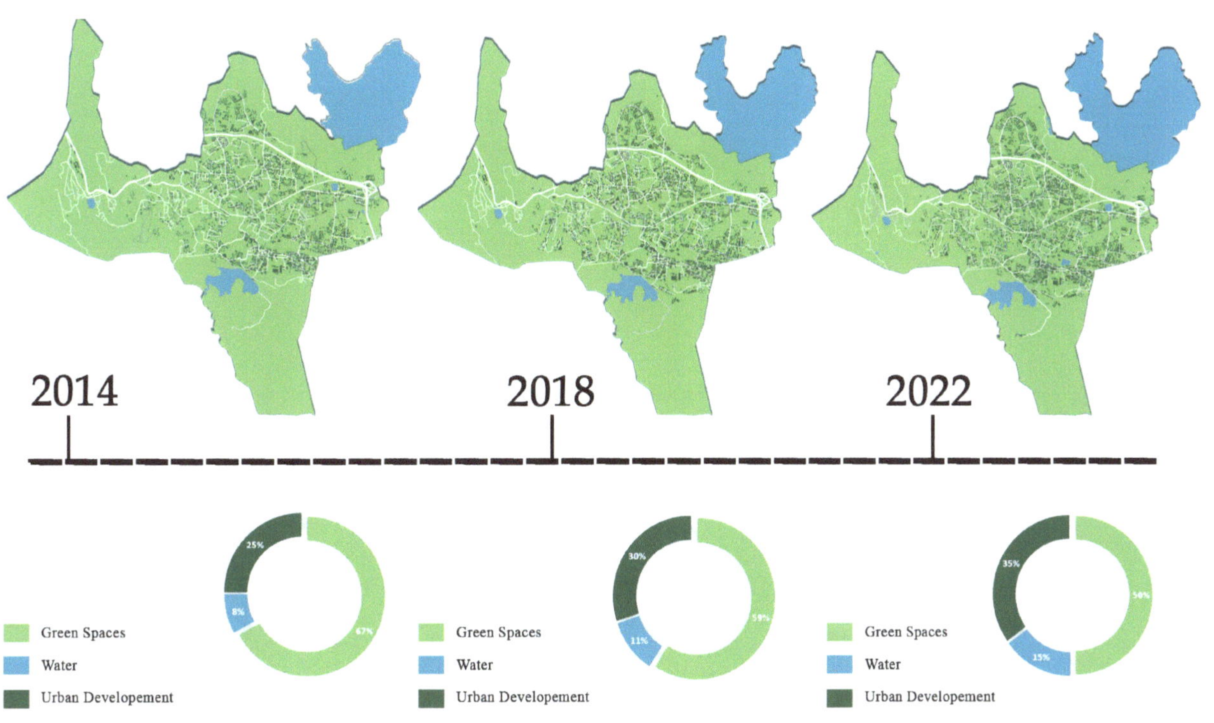

Figure 3: Evolution of land cover through the years

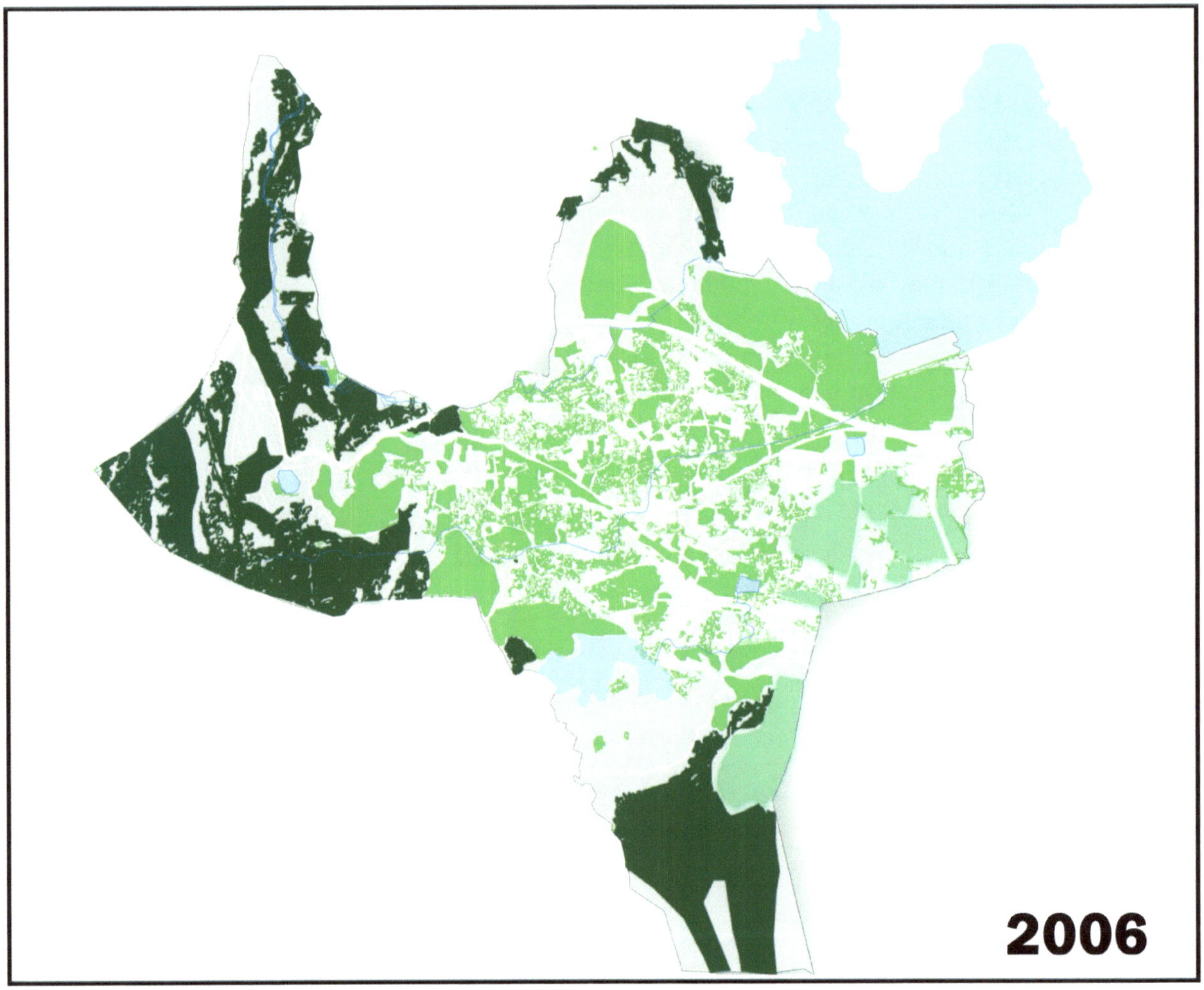

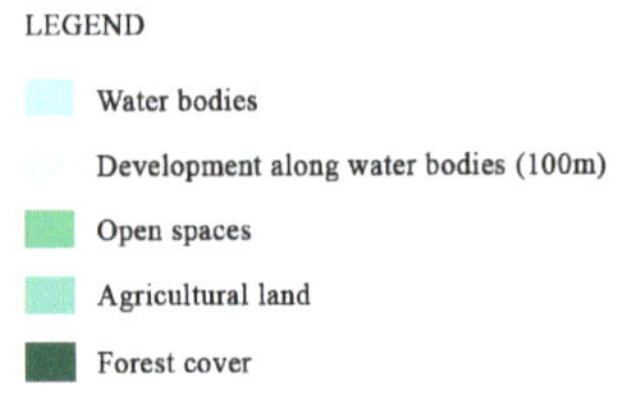

Figure 4: Tracing open spaces 2006

LEGEND

- Water bodies
- Development along water bodies (100m)
- Open spaces
- Agricultural land
- Forest cover

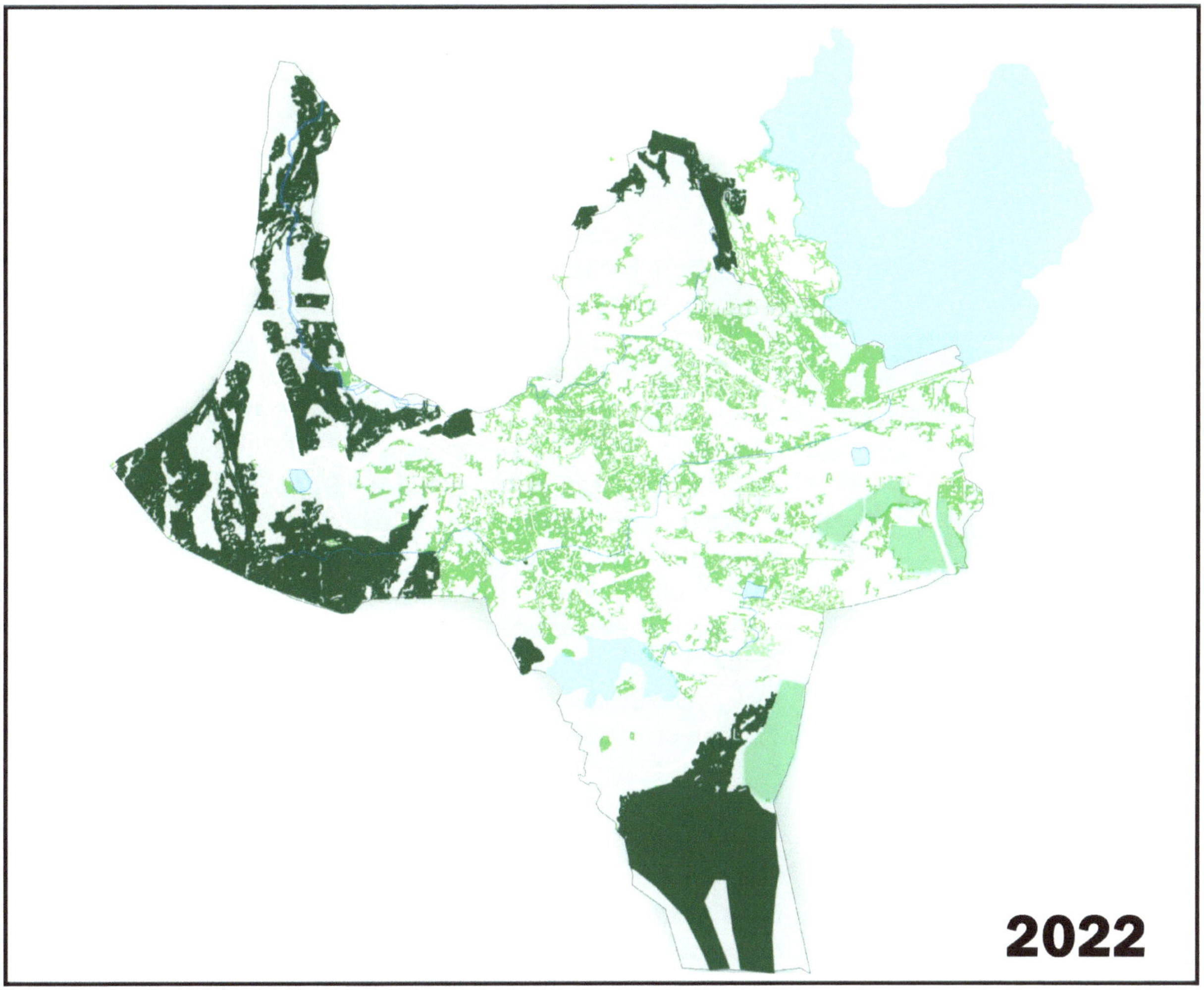

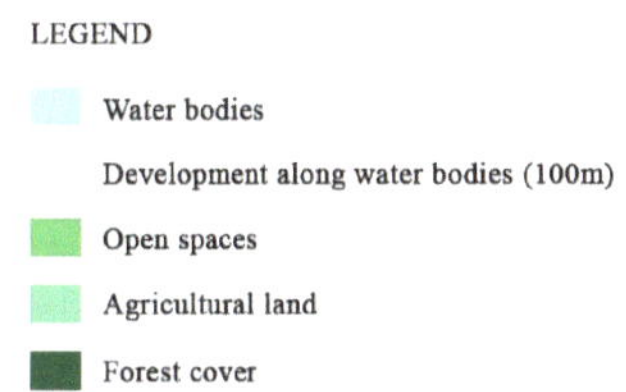

Figure 5: Tracing open spaces 2022

LEGEND

Water bodies

Development along water bodies (100m)

Open spaces

Agricultural land

Forest cover

Analysis: Developemnt along water bodies increase from 2006 to 2022. Open spaces,Agricultural land and forest area started decreasing over the period.

1.3 Peaks and Valleys: Topographical Diversity

Lonavala's topography is characterized by rolling hills, expansive plateaus, and lush greenery that thrives in the region's moderate climate. The terrain is largely shaped by the Sahyadri mountain range, with valleys and ridges that add to its varied landscape. During the monsoon season, mist drapes the valleys, creating a breathtaking visual effect, while the abundant rainfall nourishes the dense forests and rich vegetation.

One of the most notable features of Lonavala's topography is its network of lakes and reservoirs, which not only enhance the region's scenic beauty but also serve as vital water sources. Pawna Lake, for example, is surrounded by hills and is a popular site for camping and boating. Similarly, Bhushi Dam is famed for its cascading waterfalls, where water flows over rocky terrain, drawing visitors for its natural allure.

The region's diverse ecological zones are home to a variety of flora and fauna, particularly within the dense forests that cover much of the landscape. These forests contribute significantly to the local biodiversity and act as green corridors within an otherwise rapidly urbanizing area.

Lonavala's historical elements are also embedded within its topography. The Karla and Bhaja Caves, which date back to the 2nd century BCE, are rock-cut cave systems that showcase ancient Buddhist architecture and stupas. These sites, carved into the hills, stand as remnants of the region's rich cultural and religious heritage.

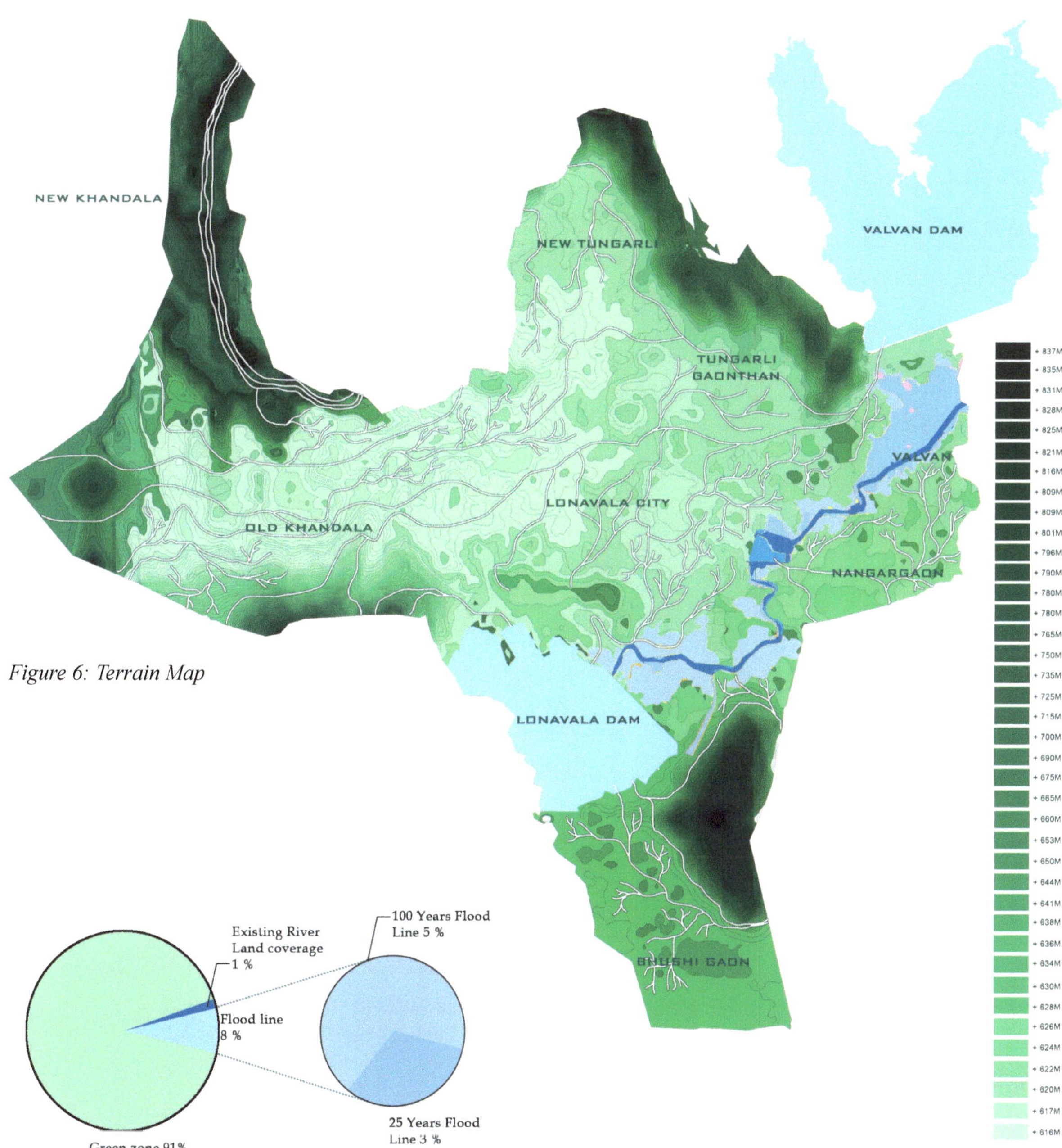

Figure 6: *Terrain Map*

Figure 7: *Topoghrapical analysis*

Image 3: Tunnels and road network carved through the controus of the region with the settlement spread over the region.

Image 4: Bhor ghat, Khandala

Image 5: Bhor ghat, Khandala -1895

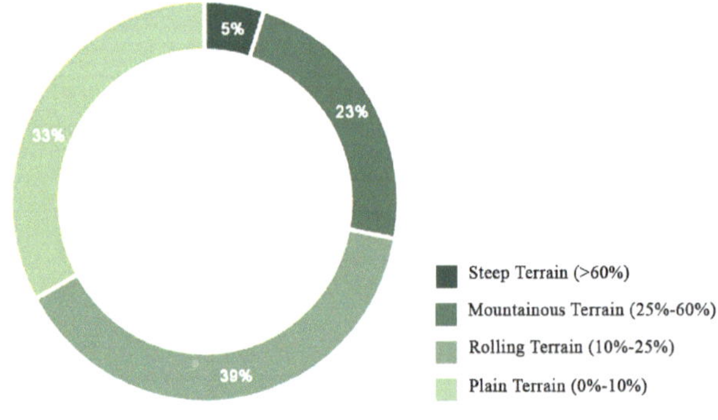

Facts:

Forest
- Max. Slope - 94.4%
- Avg Slope - 17.5% to 18.5%
City
- Max. Slope - 48.7% to 53.3%
- Avg. Slope - 5.8% to 7.4%

Figure 8: Sectional figure depicting Lonavala's landform

Kune Falls Fragment: Kune Falls is situated between the twin hills of Lonavala and Khandala, at an elevation of 622 meters above sea level, framed by the majestic Sahyadri Mountains in western Maharashtra. Ranking as the 14th tallest waterfall in India, Kune Falls features a total height of approximately 200 meters, divided into three distinct sections.

Image 6 : This is the lonavala dam ,which is a watershed area where water from the surrounding mountains is channelled into the dam . Also the dam is build on the Intrayani river which furthure flows into the city and merges with the bhima river.

1.4 UNRAVELLING THE WATERSHED

Lonavala experiences substantial rainfall during the monsoon season, which typically lasts from June to September. The region's watershed includes major rivers such as the Indrayani River and the Pavana River, both of which eventually flow into the Bhima River—a key tributary of the Krishna River.

The Western Ghats, which encompass Lonavala, serve as a natural barrier, capturing moisture-laden winds from the Arabian Sea and resulting in heavy precipitation. This abundant rainfall fosters the formation of numerous streams and rivers within the Lonavala watershed.

The water from this watershed is an essential resource for both domestic use and agricultural activities in the surrounding areas. The rivers and reservoirs created by this watershed are crucial water sources for nearby cities and towns, including Lonavala itself.

Image 7: Tata dam along the sided of Lonavala lake, 2023.

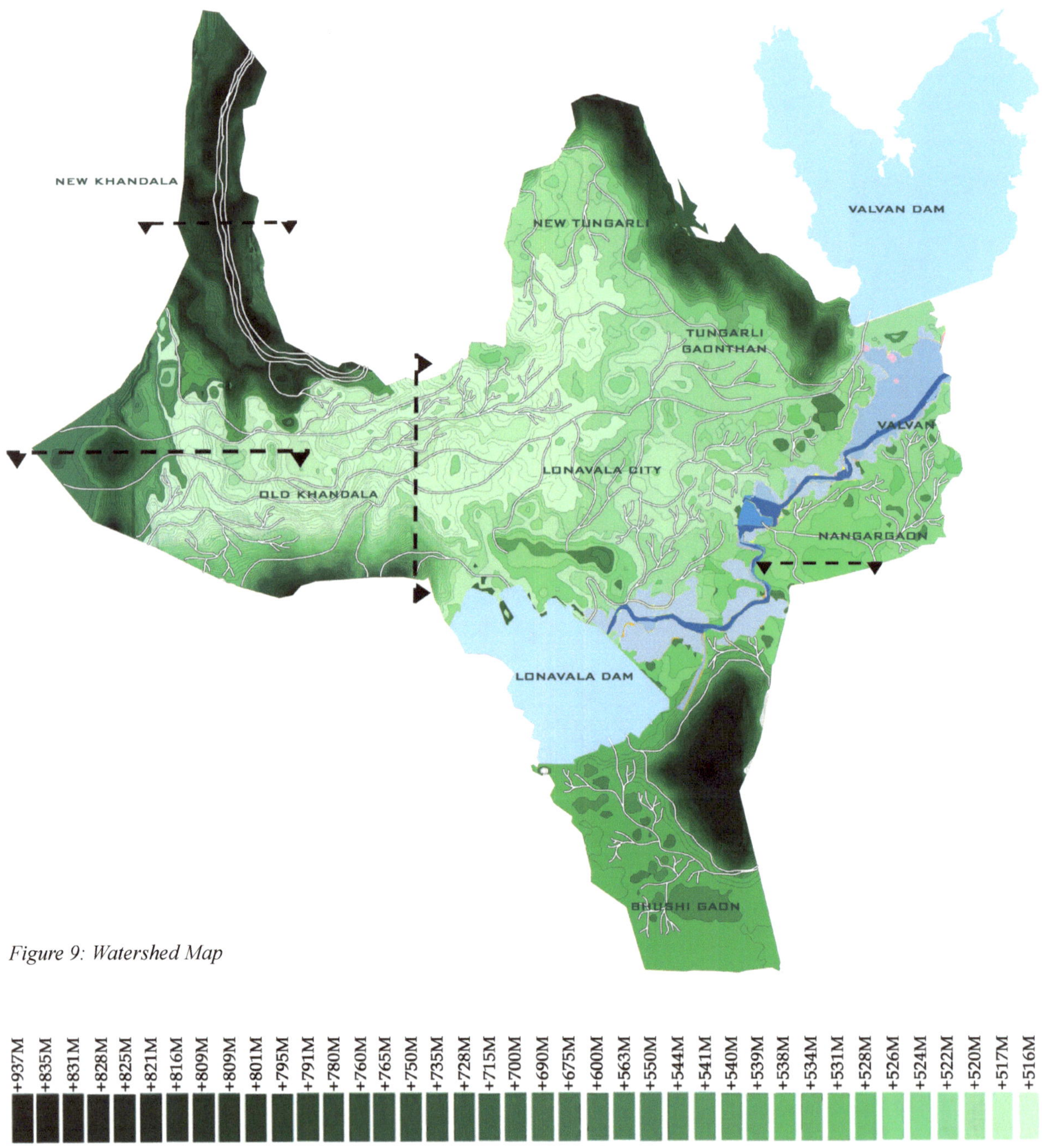

Figure 9: Watershed Map

Section AA'-
Major break in forest corridor in Khandala and lonavala region due to Mumbai and Pune Expressway.

Section BB'-
Western Ghat have Shola grasslands where the forest grows along sheltered stream beds while the grassland patches cover the intervining hill slopes.

Section CC'-
There are valleys inundated by heavy cloud cover for prolonged periods which is known to a count for profusion of species of plants. including ground, canopy, orchids, ferns and fungi.

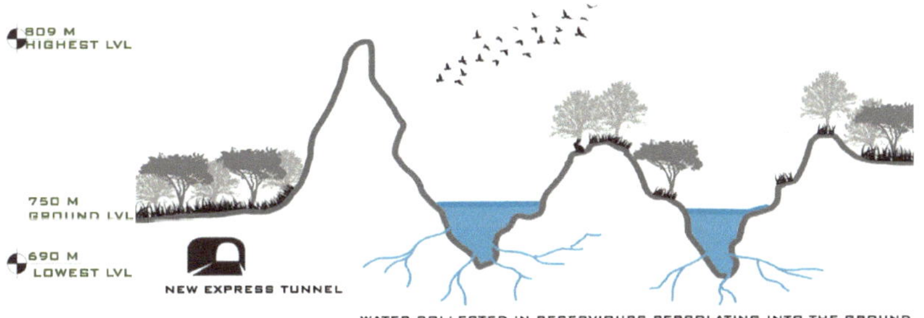

Section DD'-
Lonavala is located in the lap of Sahyadri with peaks and valleys which are suitable for dams.

Image 8: Connecting higways and expressways to Lonavala, 2023.

1.5 Wildlife Symphony: Exploring the Flora and Fauna

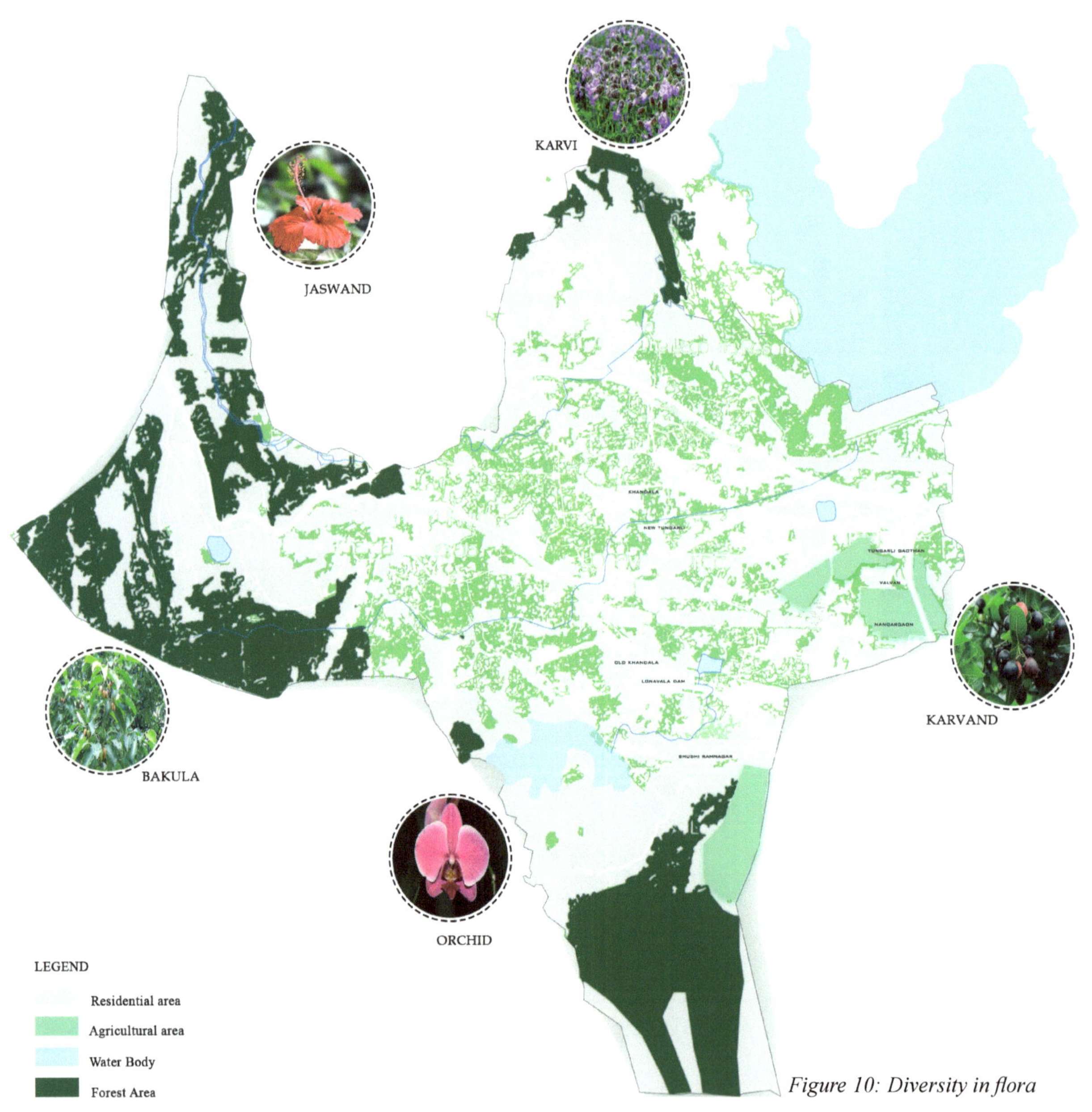

Figure 10: Diversity in flora

FLORA

The natural vegetation found in Lonavala is influenced by climate . The forest found in Lonavala is of semi evergreen forest which is found in the high rainfall areas of the Sahyadri hills sometimes called subtropical evergreen forests. The trees are evergreen because of the heavy rainfall in this region. These forests have been mostly destroyed by urbanisation and infrastructure development and occur only in patches.

Commonly found shrubs in the region

Jaswand (Hibiscus Rosa-Sinensis) The vibrant and captivating Jaswand, also known as Hibiscus, adds a splash of colour to Lonavala's gardens and landscapes.

Karvi (Strobilanthes Callosus)

Karvi is a remarkable plant that is known for its unusual blooming pattern. It flowers gregariously at intervals of several years, covering the hill sides with a spectacular display of purple blossoms. The flowering of Karvi is a natural phenomenon that attracts attention and adds a unique charm to the region.

Bakula (Mimusops Elengi)

The fragrant Bakula tree is cherished for its sweet-scented flowers and dense foliage. Bakula flowers, often used in perfumes and traditional medicine, are small and star-shaped, emitting a delightful aroma that lingers in the air. The Bakula tree's presence contributes to the sensory experience of Lonavala's natural environment.

Orchid (Orchidaceae)

Orchids, a diverse and captivating group of flowering plants, can be found in various habitats around Lonavala. These plants are renowned for their intricate and elegant flowers that come in a stunning array of shapes, colors, and sizes.

Karvand (Carissa Congesta) The Karvand, also known as Carissa or Karonda, is a fruit-bearing shrub that contributes to both the flora and local cuisine of Lonavala. The small, round Karvand fruits are used to make jams, jellies, and pickles. The plant's hardy nature allows it to thrive in various conditions, making it a resilient part of Lonavala's plant community.

Image 9: Khandala Ghat

Image 10: Ryewood Park

Image 11: Likely seen dense vegetation
source : mention the source hear

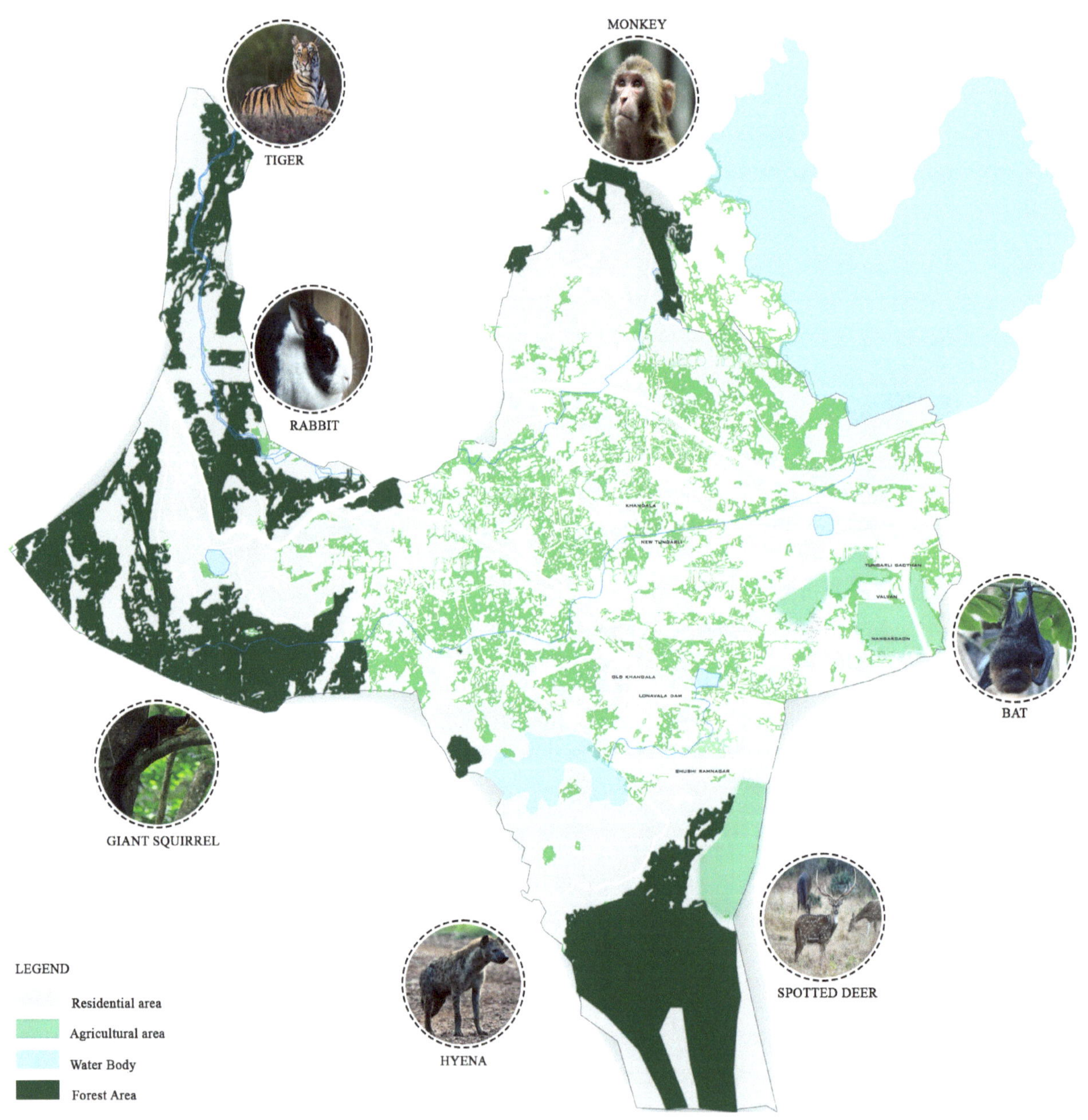

Figure 11: Diversity in fauna

FAUNA

Lonavala maintain regional biodiversity. The Fauna includes following animals.

Tiger (Panthera Tigris)

The elusive and majestic tiger, an apex predator, can be found in the forests surrounding Lonavala. Tigers play a crucial role in maintaining the ecological balance of the region by controlling prey populations. Although sightings are rare, their presence adds to the wilderness allure of Lonavala.

Monkey (Macaca sp.)

Various species of monkeys, such as the rhesus macaque and the bonnet macaque, inhabit the forests and urban fringes of Lonavala. These intelligent and social animals are often seen swinging through the trees and interacting with each other, adding a lively touch to the landscape.

Rabbit (Oryctolagus Cuniculus)

The black and white rabbit, also known as the European rabbit, can be spotted in the grasslands and open areas of Lonavala. These small mammals are known for their distinctive coat pattern and are important herbivores in the local ecosystem.

Giant Squirrel (Ratufa Indica)

Lonavala's forests are also home to the impressive giant squirrel. With its vibrant coat and large size, this arboreal creature is known for its graceful leaps between trees and its vital role in seed dispersal.

Spotted Hyena (Crocuta Crocuta)

The spotted hyena is a carnivorous predator that can be found in the wilderness of Lonavala.

Spotted Deer (Axis Axis)

Also known as chital, the spotted deer is a common sight in the forests of Lonavala. These graceful herbivores graze in open areas and are prey for predators like tigers and hyenas.

Bat (Chiroptera)

Lonavala is home to a diverse range of bat species, which are crucial for pollination.

Image 12: Central core of Lonavala

2. Urban Morphology

Strategically situated along the Mumbai-Pune corridor, Lonavala's geographical location has been instrumental in shaping its historical and contemporary significance. The town's accessibility via well-developed road networks and the Mumbai-Pune Expressway has made it a favored destination for weekend getaways and short trips from both metropolitan centers. This easy connectivity has not only contributed to the flourishing tourism industry but has also paved the way for real estate development, with many people investing in vacation homes and second properties in the area.As a result, the city of Lonavala is showing a remarkable growth along its fringes.

The allure of Lonavala lies in its verdant landscapes, majestic hills, cascading waterfalls, and the soothing climate, making it a pleasant escape from the scorching heat of nearby coastal regions. Tourists and nature enthusiasts find solace in the lush greenery, picturesque viewpoints like Tiger's Leap and Lion's Point, and the serene ambiance of Bhushi Dam. Additionally, the historical Karla and Bhaja Caves showcase the region's rich cultural heritage and attract historical aficionados.

The town's proximity to Mumbai and Pune also enhances its appeal as a recreational hub, with a plethora of adventure activities and outdoor pursuits available to visitors. From trekking and hiking in the surrounding Sahyadri mountains to exploring ancient forts like Rajmachi, there is an abundance of experiences to cater to varying interests.

Lonavala's development over the years has seen the establishment of numerous hotels, resorts, restaurants, and recreational facilities, catering to the growing influx of tourists. Despite this progress, efforts are made to preserve the region's natural beauty and ecological balance, as it remains an essential part of its charm and attraction

2.1 JOURNEY OF URBAN GROWTH

Ancient History:
Lonavala's history can be traced back to ancient times, and it has been mentioned in various historical texts. Lonavala boasts some ancient architectural wonders, including the Karla Caves and the Bhaja Caves. These Buddhist rock-cut caves date back to the 2nd century BC and showcase intricate carvings and sculptures. For a prolong period, it was known as a small settlement with agricultural activities, and also for its natural beauty and serene surroundings.

British Colonial Period:
During the British colonial era, Lonavala gained importance due to its strategic location on the trade route connecting Mumbai (then Bombay) to Pune. The British recognized its potential as a hill station and established several bungalows and retreats for officers and soldiers seeking respite from the heat of the coastal regions.

Image 13: Then Karli caves presently known as the famous Karla caves.

Post-Independence Era:
After India gained independence in 1947, Lonavala witnessed further development in terms of infrastructure, education, and healthcare facilities. As urbanization and industrialization increased in nearby cities, the demand for weekend getaways and second homes also contributed to the town's growth.

19th Century:
The development of Lonavala as a hill station accelerated in the 19th century when the Mumbai-Pune railway line was constructed, and a railway station was established in the town. This enhanced accessibility, leading to an increase in tourism and further development of infrastructure.

20th Century:
In the 20th century, Lonavala continued to attract tourists, and its popularity as a weekend getaway and a holiday destination grew among people from nearby urban centers like Mumbai and Pune. The town saw the establishment of more hotels, lodges, and recreational facilities to cater to the rising number of visitors.

21st Century:
In recent decades, Lonavala has transformed into a bustling town with a mix of residential, commercial, and tourist-centric developments. The real estate sector has seen significant growth, with more people investing in properties for vacation homes or retirement living.

Image 14: Present day urban fabric showing recreational spaces- Rama Krishna hotel and Della adventures

2.2 THE URBAN VEINS: MAJOR ROUTES AND NODES

The proximity from two major cities has been instrumental in development of good connectivity in the form of roads and railways. The ever increasing tourist footfall over weekends has questioned the capacity of the present day infrastructure paving a way to further development of routes. Once such route being the 'missing link road'.

The 'missing link road' is set to revolutionize travel between Mumbai and Pune, offering a groundbreaking solution to reduce travel time and enhance safety on the Mumbai-Pune Expressway. This landmark project utilizes cutting-edge technology used globally, boasting the world's widest tunnel, with a width of 23.75 meters.

Image 15: A network of major roads,railway line connecting to Lonavala.

LEGEND
- NH4 Mumbai Pune Express way
- NH48 Old Mumbai Pune Express way
- Railway line
- Underground by-pass from Sighgad Institute to Khalapur
- Flyover above the railway track
- Waterbody
- Hill area
- Lonavala railway station

Image 16: Complex road networks and interconnectivity at Center Point-an important node in lonavala

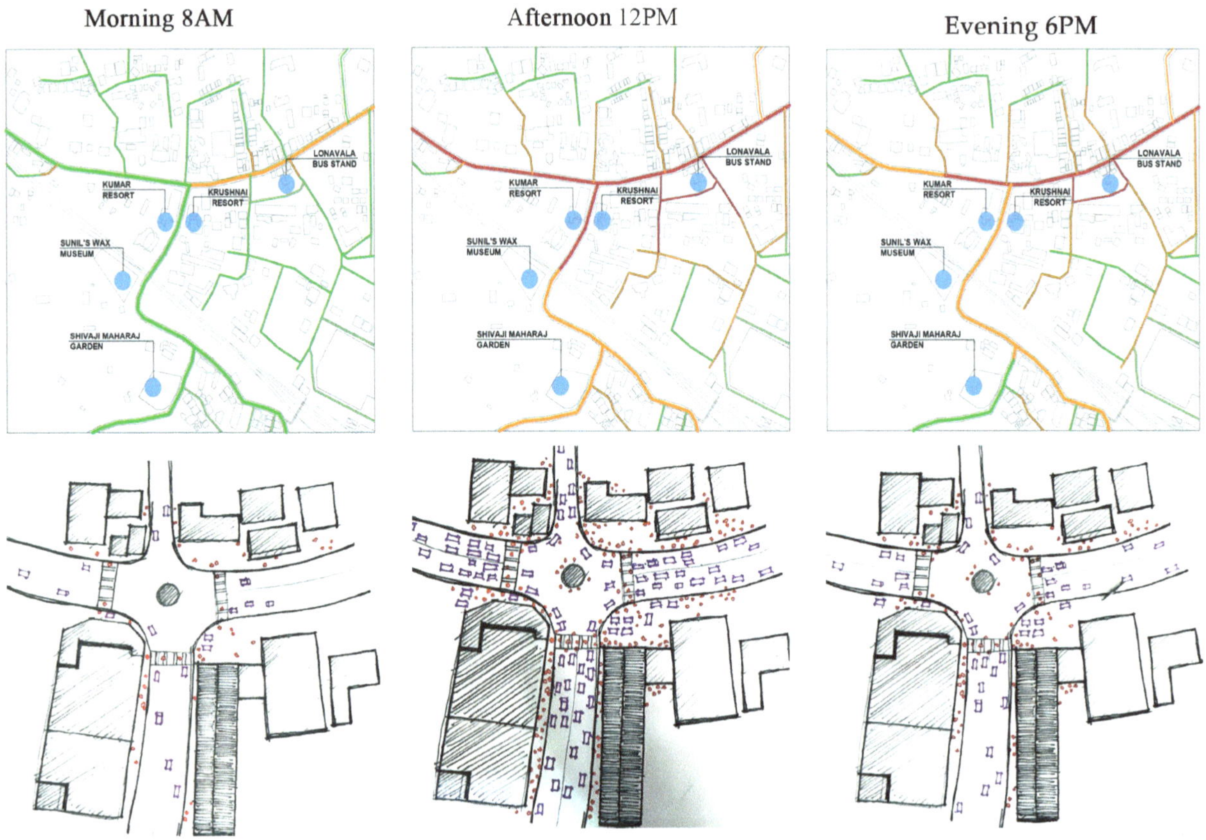

Morning 8AM   **Afternoon** 12PM   **Evening 6PM**

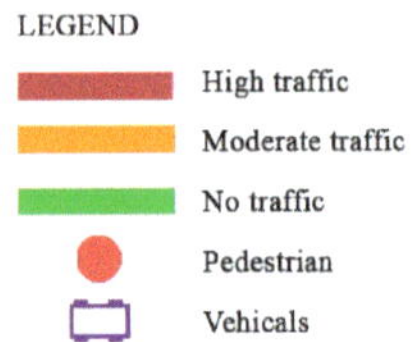

Figure 12: Nodal traffic analysis at Kumar resort junction

Image 17: Kumar resort junction at 8 AM in the morning.

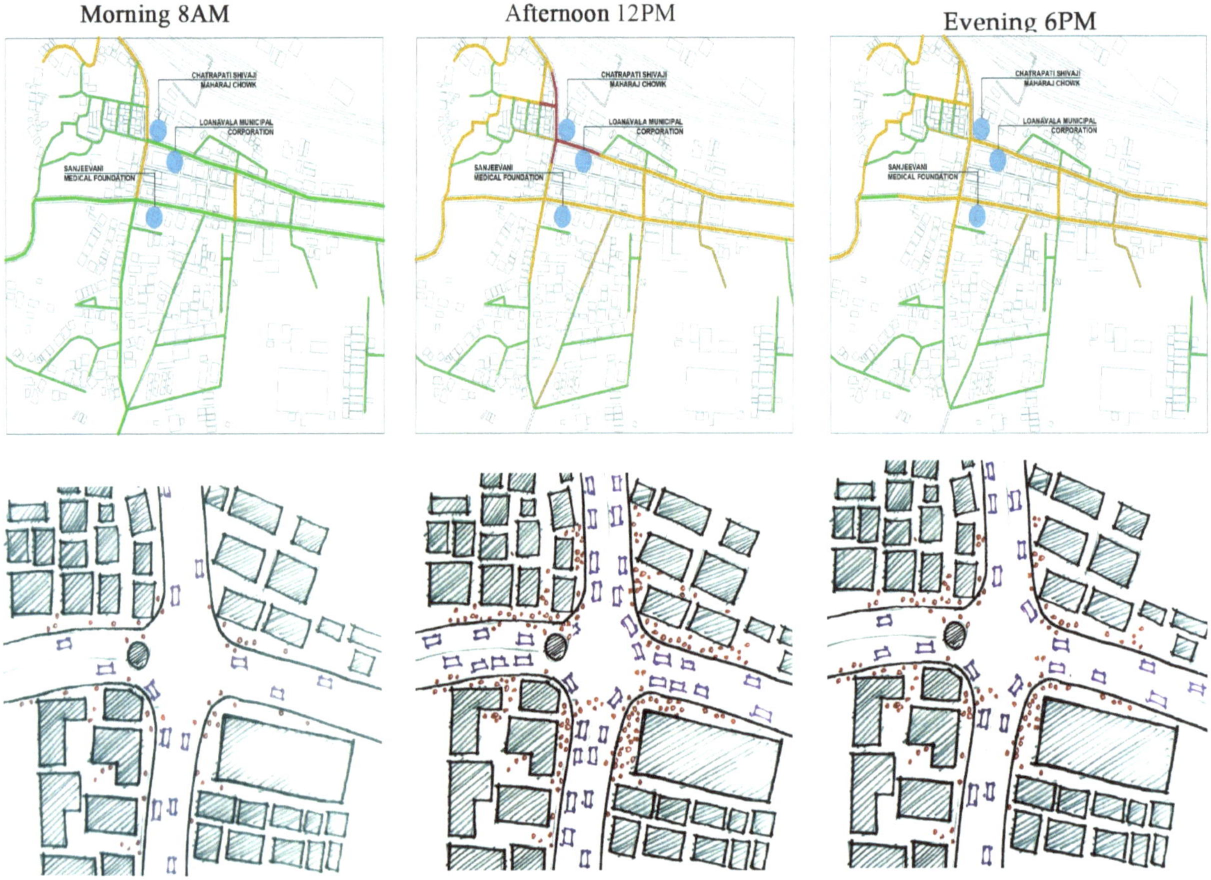

Morning 8AM　　**Afternoon 12PM**　　**Evening 6PM**

Figure 13: Nodal traffic analysis at Shivaji Maharaj Chowk

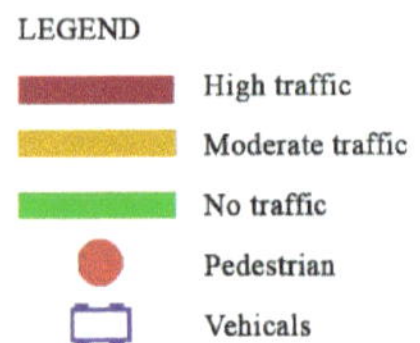

Image 18: Shivaji Maharaj chowk at 8 AM in the morning.

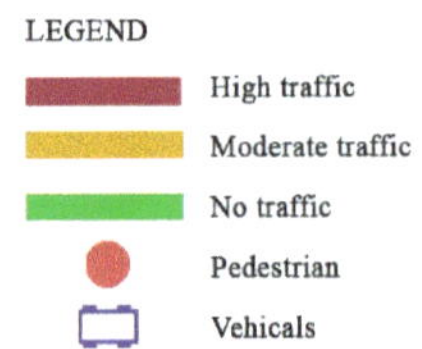

Figure 14: Nodal traffic analysis at Lohgad Udyan

Image 19: Lohgad Udyan at 8 AM in the morning.

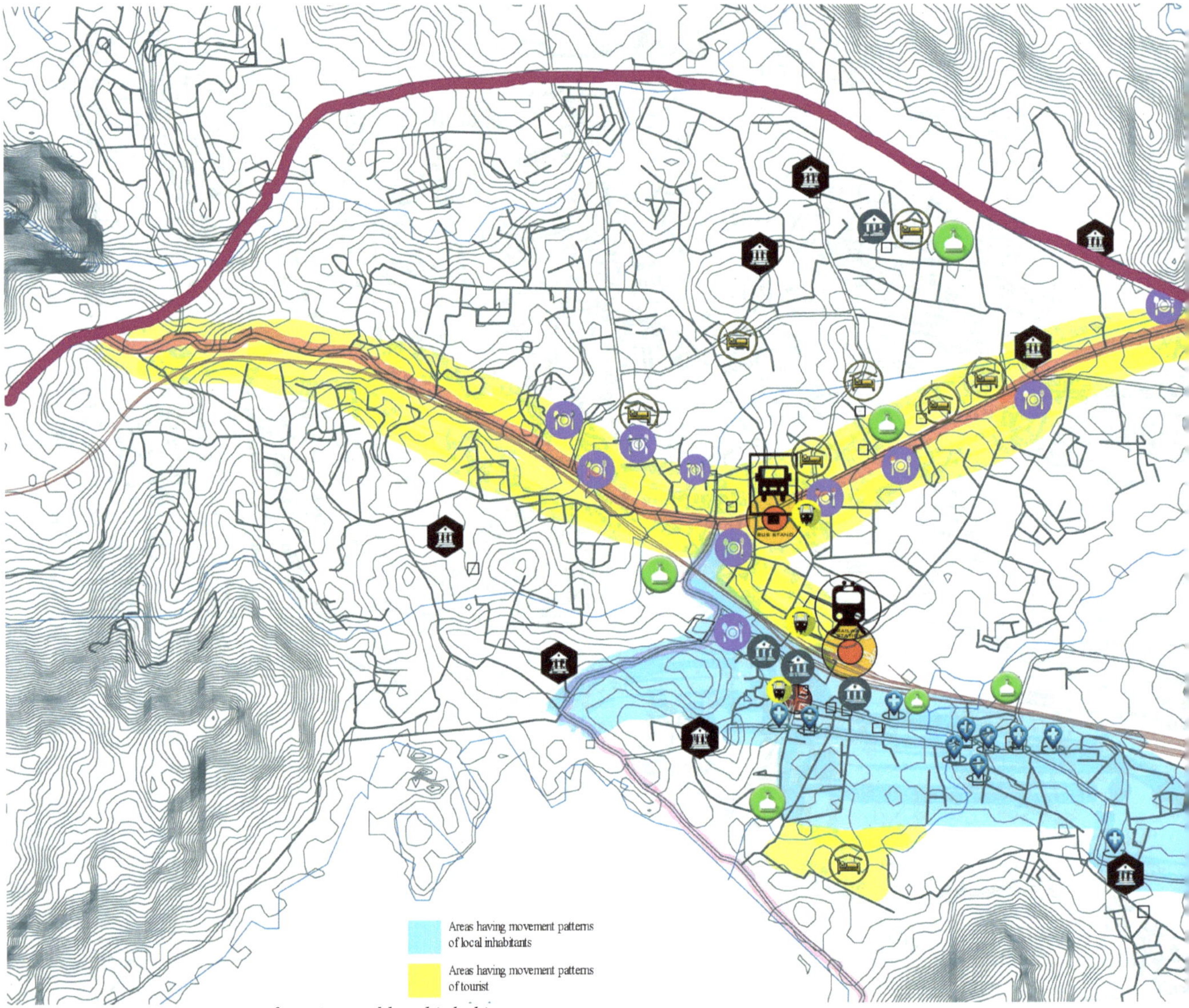

Image 20: Movement patterns of tourist and local inhabitants.

Waterfall

Bhushi dam

Lohagad

Local scenery

Local food

Image 21: Famous tourist attractions.

2.3 SEASONS OF SERENITY: VISITING ENCHANTING LONAVALA

This popular weekend destination draws in over 61,846 tourists every weekend. Also, there are over 1600 tourists visiting more than 5000 privately owned bungalows & holiday homes every week.

The 92 hotels & resorts in Lonavala, with a capacity of over 2,446 rooms enjoy 59% occupancy rate through the year. About 78% of these tourists come here car.

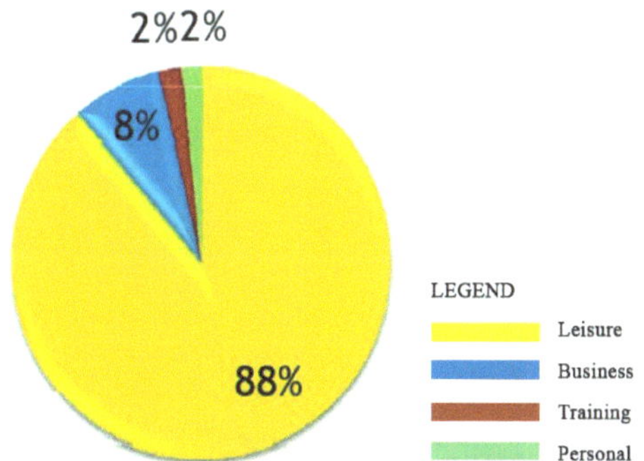

Figure 15: Key reasons to visit

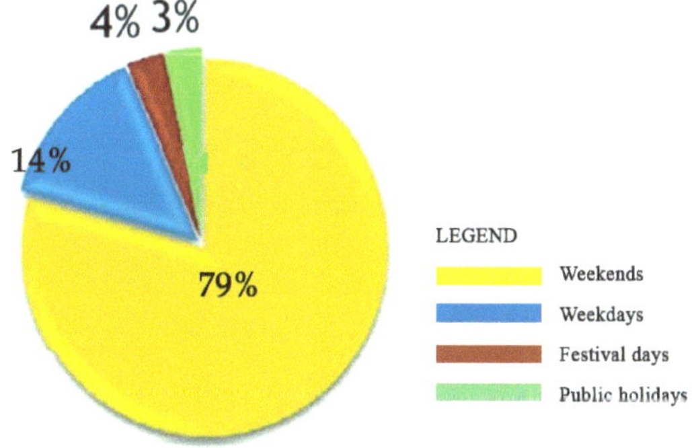

Figure 16: Prefered day to visit

Image 22 : Reminiscent of street shopping experienced during Friday market, 2023.

Image 23: A perspective of Friday market street depicting the essence of a vibrant shopping experience, 2023.

Image 24 : A perspective of Friday market at Purandar ground depicting the essence of a vibrant shopping experience, 2013.

2.4 Urban Fabric

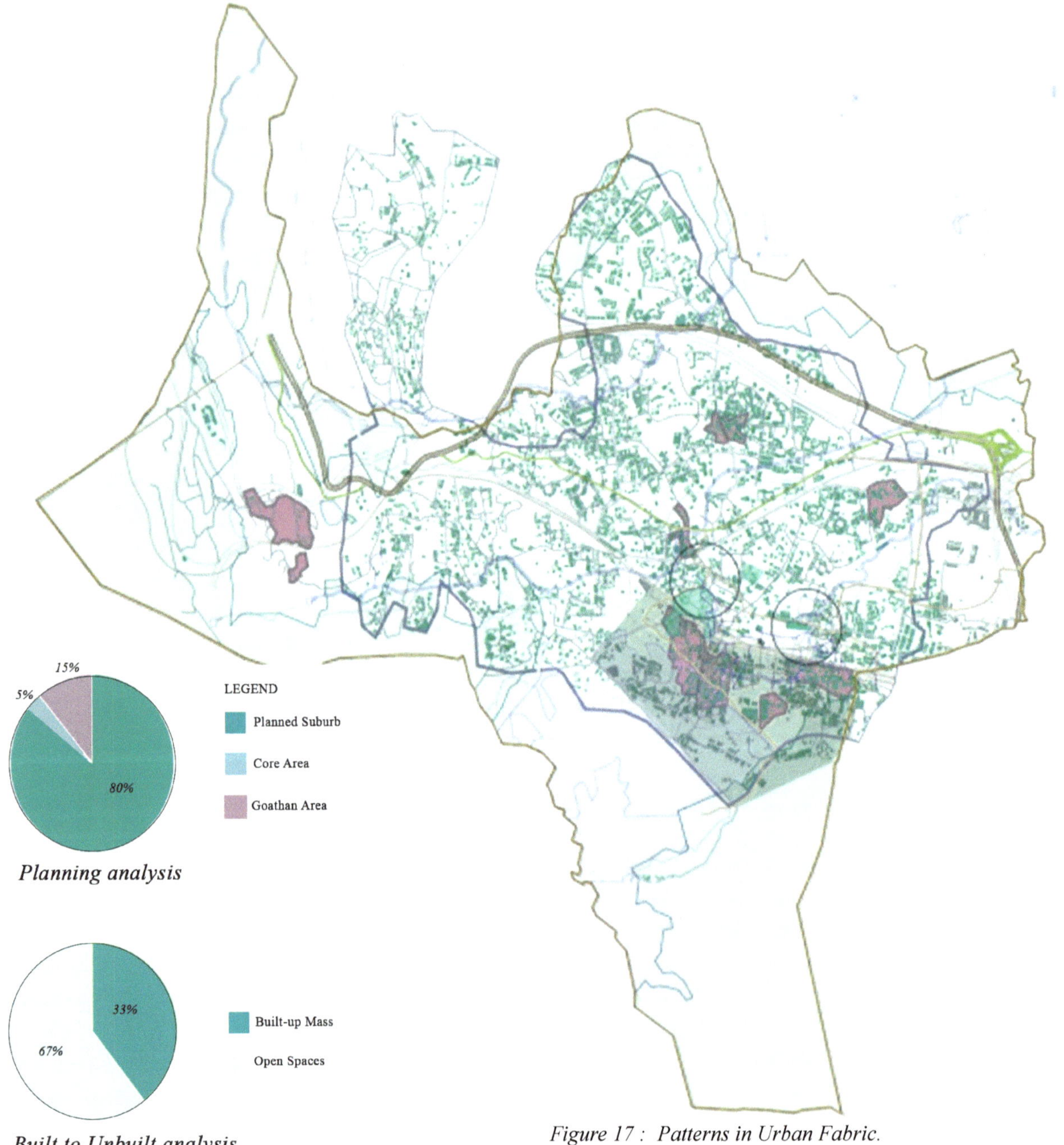

Figure 17 : Patterns in Urban Fabric.
(Planning, Built to Unbuilt analysis)

Image 25: Jaichand chowk - A Prominent node.

The urban fabric of lonavala shows a progressive blend of vernacular and modern architecture, tracing it's development over untraceable period of time. The present day architecture of the city represents modern frame structures and ancient load bearing structures exhibiting the usage of modern and ancient materials. In an attempt to nurture the goodness of weather and scenic beauty, the city of lonavala shows traces of passive space design strategies being adopted in built forms. Within the complexity of the growing urban fabric, devels a unique character of the city which can be understood by the built volume and spaces thus formed inbetween

Figure 18: Building Typology analysis.

Image 26: From Shivaji Maharaj chowk towards Bhangarwadi

Image 27: From Shivaji Maharaj chowk towards Gaonthan

Figure 19: Urban Growth Pattern- year 2007

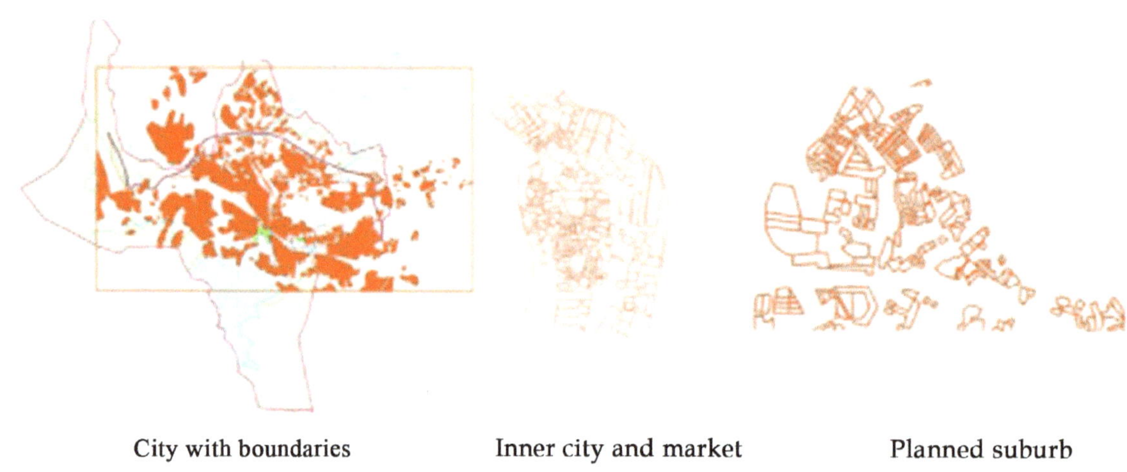

Figure 20: Urban Growth Pattern- year 2013

Figure 21: Urban Growth Pattern- year 2017

Figure 22: Urban Growth Pattern- year 2022

Image 28: View of Khandala, a small hill station in the state of Maharashtra, 1860s.

Image 29 : A waterfall near Lonavala lake.

3. Economical Mapping
Socio- economical dynamics

Lonavala, situated between Mumbai and Pune, displays unique social and economic characteristics shaped by its built environment. Architecturally, the town combines traditional Maharashtrian vernacular structures with contemporary developments driven by tourism. Traditional homes, with sloping roofs and locally sourced materials, respond to the region's climate, while modern resorts and commercial establishments cater to the growing demand from tourists and seasonal residents. This blend creates an urban fabric that reflects Lonavala's evolving socio-economic landscape.

Tourism is the primary economic driver, influencing land use and spatial organization. Key hubs like Shivaji Maharaj Chowk, the railway station, and Lonavala Market are centers of commercial activity, supporting a mix of retail, dining, and lodging. Public spaces such as Raywood Park and attractions like Bhushi Dam enhance the town's green infrastructure, meeting recreational needs while preserving its ecological character.

Economically, Lonavala relies on the hospitality and service sectors, with hotels and restaurants contributing significantly to local revenue. The town is also known for its chikki production, blending cultural heritage with commerce. However, the seasonal influx of tourists strains infrastructure, creating challenges in traffic management, parking, and public services for urban planners and architects.

Recent years have seen an expansion of Lonavala's real estate market, leading to the development of gated communities, luxury villas, and private enclaves for affluent individuals from nearby urban areas. This trend often results in the repurposing of agricultural land and open spaces, contributing to urban sprawl.

3.1 Ancient Socio-Economic Tapestry

Lonavala's geographical advantage lies in its placement along the Bhor Ghat, an ancient and vital mountain pass that facilitated trade between the coastal regions and the Deccan Plateau. This pass was an essential part of the trade networks that connected the port city of Mumbai to the interior markets of Pune and beyond, enabling the movement of goods such as spices, textiles, and agricultural products. The economic activity generated by these trade routes has historically fostered ancillary industries such as warehousing, local markets, and transportation services.

During the British colonial era, the development of the Mumbai-Pune railway line, passing directly through Lonavala, further solidified its importance as a trade and transportation hub. The construction of the Bhor Ghat Railway Incline in the mid-19th century improved the efficiency of trade flows, enhancing the connectivity between the port of Mumbai and inland markets. This railway line continues to be a vital infrastructure, facilitating both passenger and freight movement, and has significantly contributed to the region's economic development.

In the contemporary context, Lonavala's role as part of the Mumbai-Pune Expressway has facilitates rapid vehicular movement, reducing transportation time for goods and bolstering trade efficiency between the two major economic centers, supporting the flow of goods in a highly competitive industrial landscape. Furthermore, it's connectivity to both National Highway 48 and the rail freight network facilitates economic activity related to the transportation of industrial goods, raw materials, and consumer products. This has attracted investment in real estate, tourism, and allied services, adding to the diversification of Lonavala's economic base.

3.2 Heritage and Trade Networks

Lonavala and its surrounding region boast a wealth of historical sites that offer a glimpse into the area's rich cultural and architectural heritage.

The region is also home to several significant early historic and rock-cut cave sites. The Karla and Bhaja Caves, dating back to the 2nd century BCE, are among the most remarkable rock-cut cave complexes in India. These ancient Buddhist sites are renowned for their stupas, chaityas, viharas and intricate carvings, reflecting the architectural prowess and religious significance of early Buddhist communities.

These sites, collectively, offer not only a deep connection to the region's past but also serve as a testament to the architectural ingenuity and cultural richness that have shaped Lonavala's identity.

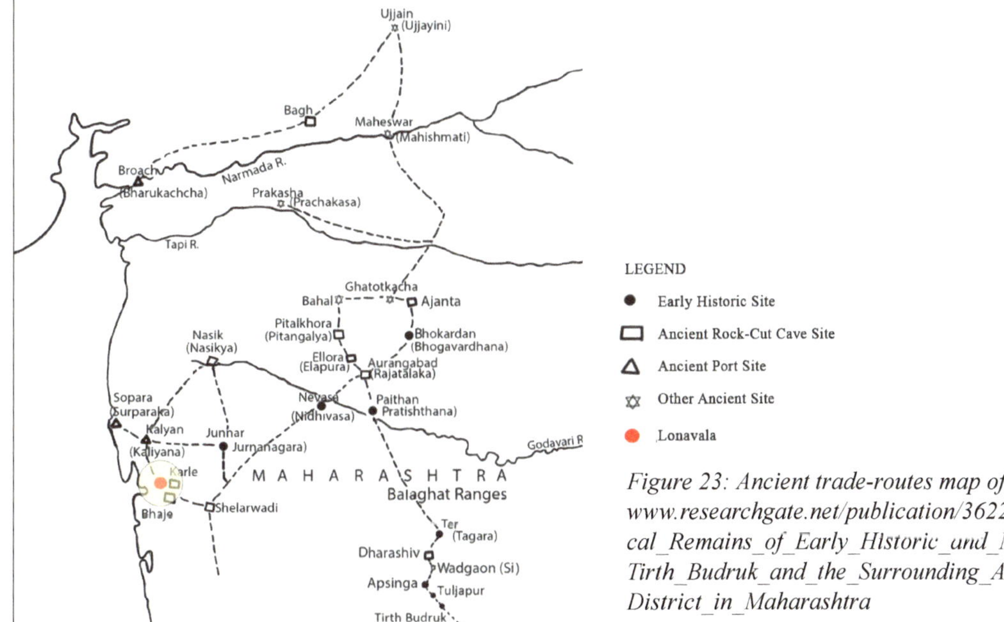

Figure 23: Ancient trade-routes map of western Indiahttps:// www.researchgate.net/publication/362213893_Archaeological_Remains_of_Early_Historic_and_Medieval_Period_at_ Tirth_Budruk_and_the_Surrounding_Area_of_Osmanabad_ District_in_Maharashtra

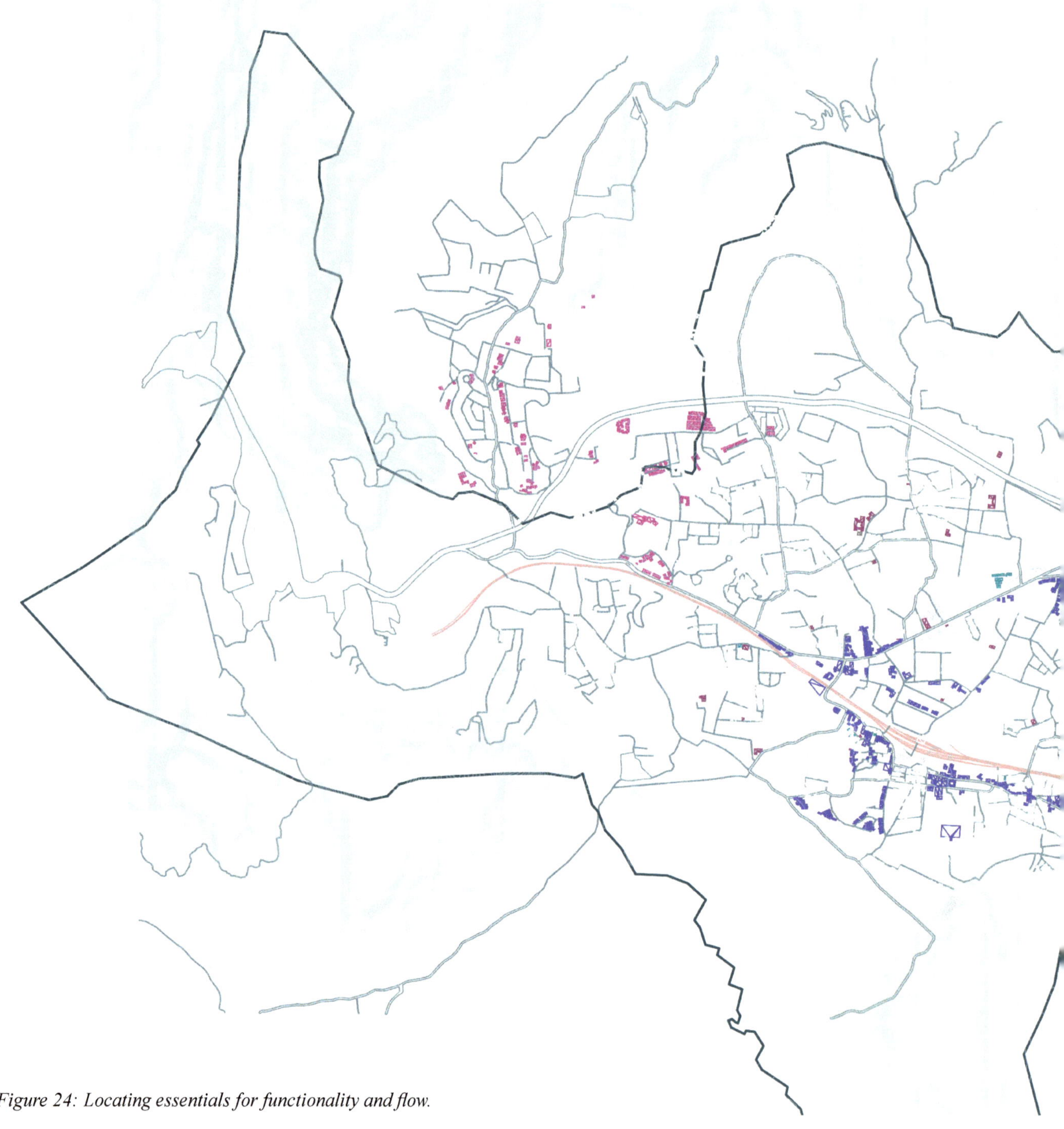

Figure 24: Locating essentials for functionality and flow.

3.3 Locating Essentials

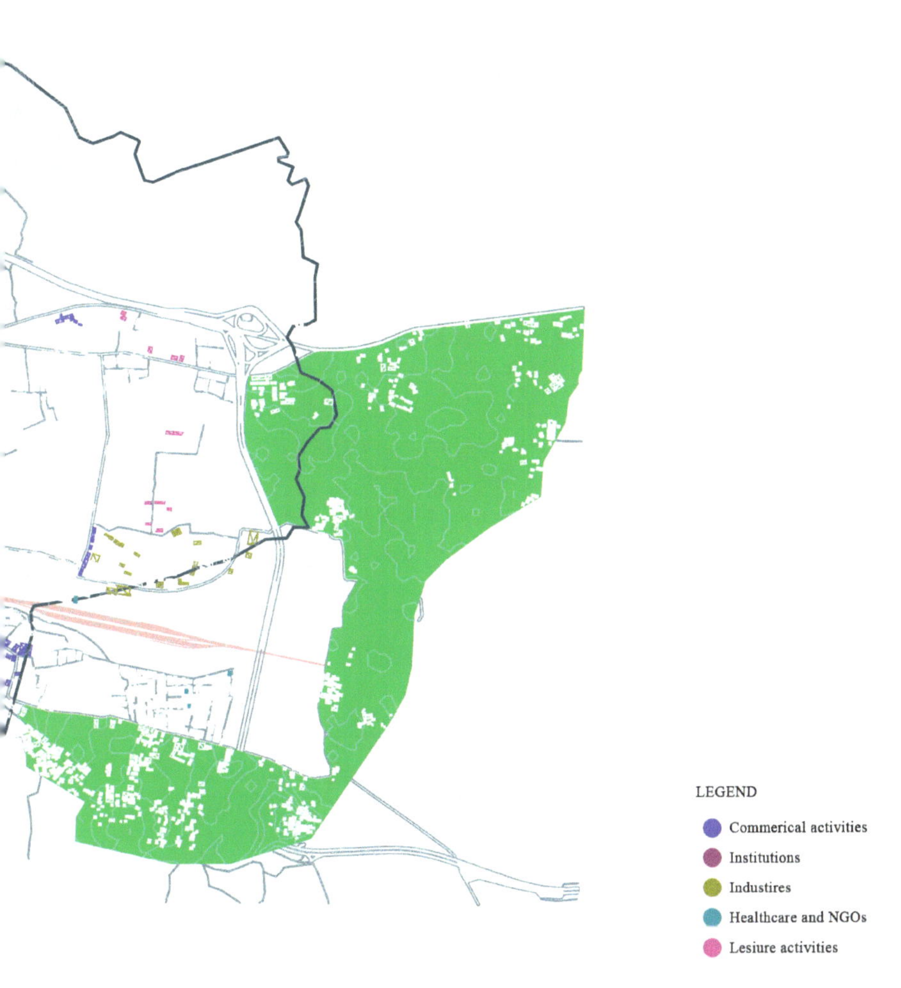

Commericals activities

Tourist-centric shops in Lonavala predominantly include chikki shops. These shops offer a wide range sweets, fudges and jellys. Tourists often flock to these shops to purchase mementos and keepsakes to cherish their experiences in Lonavala

Institutions

The town is home to a diverse range of schools that cater to the educational needs of both the local population and the influx of residents seeking a serene environment for their children's education.

Industries

One of the prominent industries in Lonavala is the manufacturing sector. The town hosts various small and medium-sized manufacturing units engaged in the production of goods such as textiles, garments, and leather products. Additionally, some factories focus on the manufacturing of food and beverages, including confectioneries and processed foods, to cater to both the local market and tourists.

Image 30: Narayani Dham
Source:https://www.bing.com/images

Image 31: Waterfall near Lonavala lake

Healthcare and NGOS

The healthcare infrastructure in Lonavala includes government-run healthcare centers, private clinics, hospitals, and medical dispensaries. These facilities offer a variety of medical services, ranging from general consultations to specialized treatments, catering to different healthcare needs.

Lesiure activities

To cater to the increasing demand for accommodation, the town offers a diverse range of resorts that provide a luxurious and tranquil escape amidst nature.These resorts are strategically situated to provide breathtaking views of the surrounding mountains, valleys, and lush greenery, creating a serene and rejuvenating ambiance for guests.A highlight of resorts in Lonavala is the inclusion of recreational facilities and activities that enhance the guest experience
The focus on hospitality and guest satisfaction is evident in the personalized services offered by the resorts

Image 32: Sterling hotel, Lonavala

Image 33: Meher villa resort, Lonavala

Image 34 : Maganlal and sons chikki shop -2023

Image 35 : Maganlal and sons chikki shop -

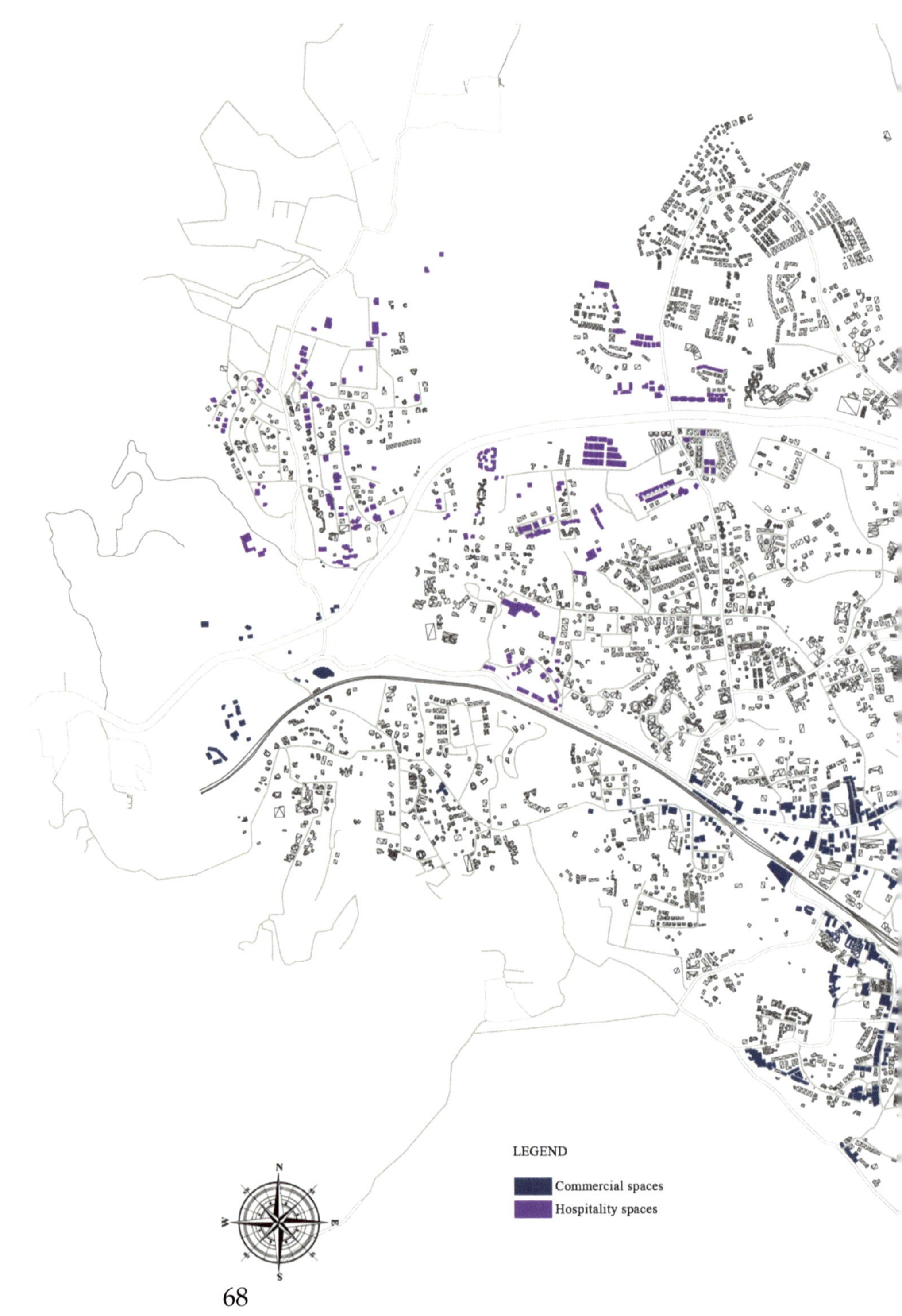

LEGEND
Commercial spaces
Hospitality spaces

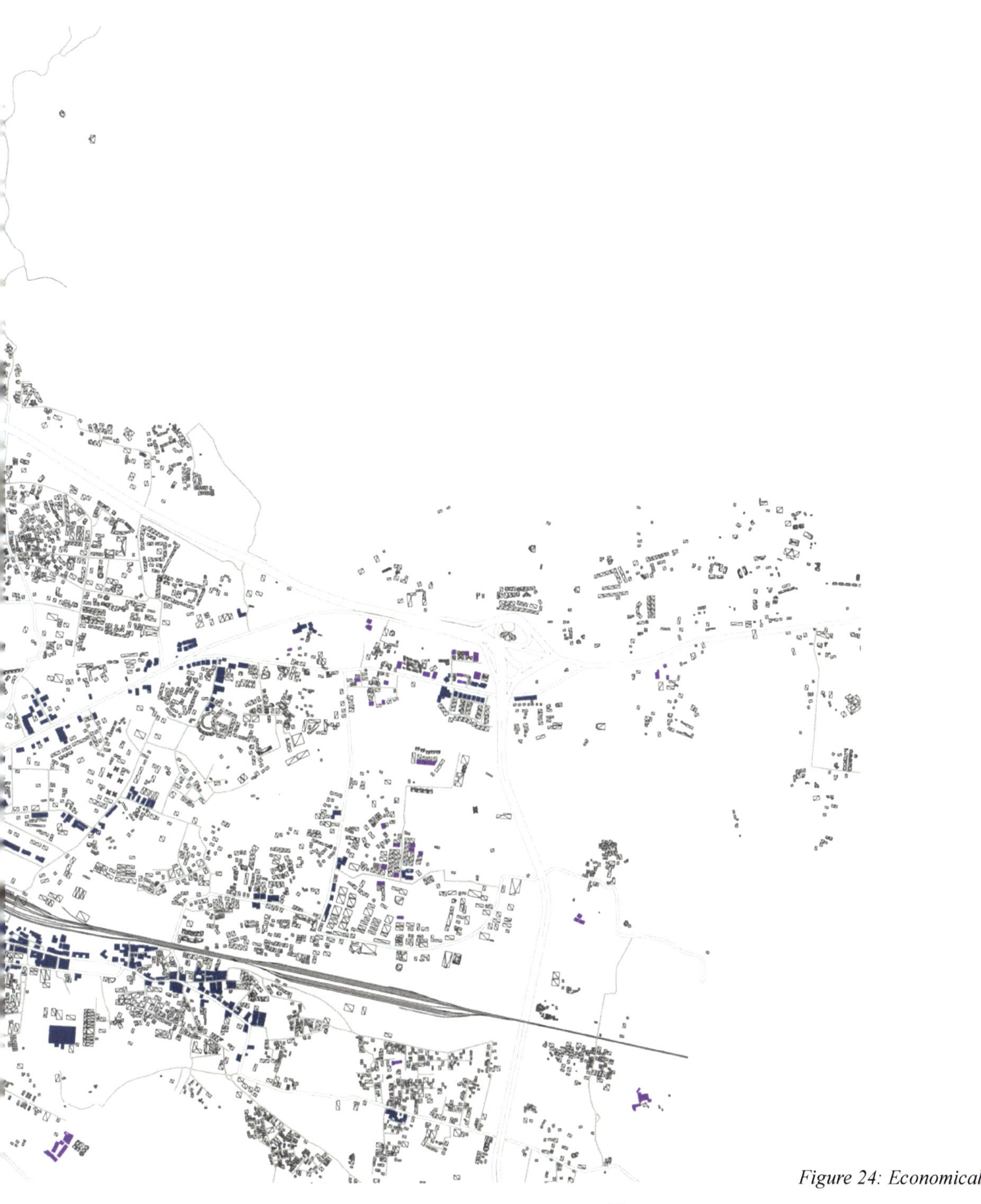

Figure 24: Economical mapping

3.4 Economic Pulse of Lonavala

Lonavala's economy is predominantly driven by its thriving tourism industry, which forms the backbone of local economic activities. The town benefits from its accessibility and scenic appeal, attracting domestic and international visitors.

Lonavala is renowned for its production of chikki, a traditional confection made from jaggery and nuts, which has become an iconic product synonymous with the town. The chikki industry, deeply rooted in local culture, has expanded significantly to meet the demands of the tourist market. This industry not only supports local artisans and small-scale manufacturers but also contributes to the regional identity and economy through retail sales and exports.

The town also experiences seasonal economic activities tied to local festivals and public holidays, which see a surge in both visitor numbers and consumption patterns. These periods of heightened activity drive demand in ancillary sectors such as food and beverage services, transportation, and leisure activities. Moreover, Lonavala's real estate market has seen substantial growth, particularly in developing second homes and vacation properties. This has led to increased investment in construction and real estate services, contributing to the overall economic diversification of the region.

Image 36 :Bhaji Mandai

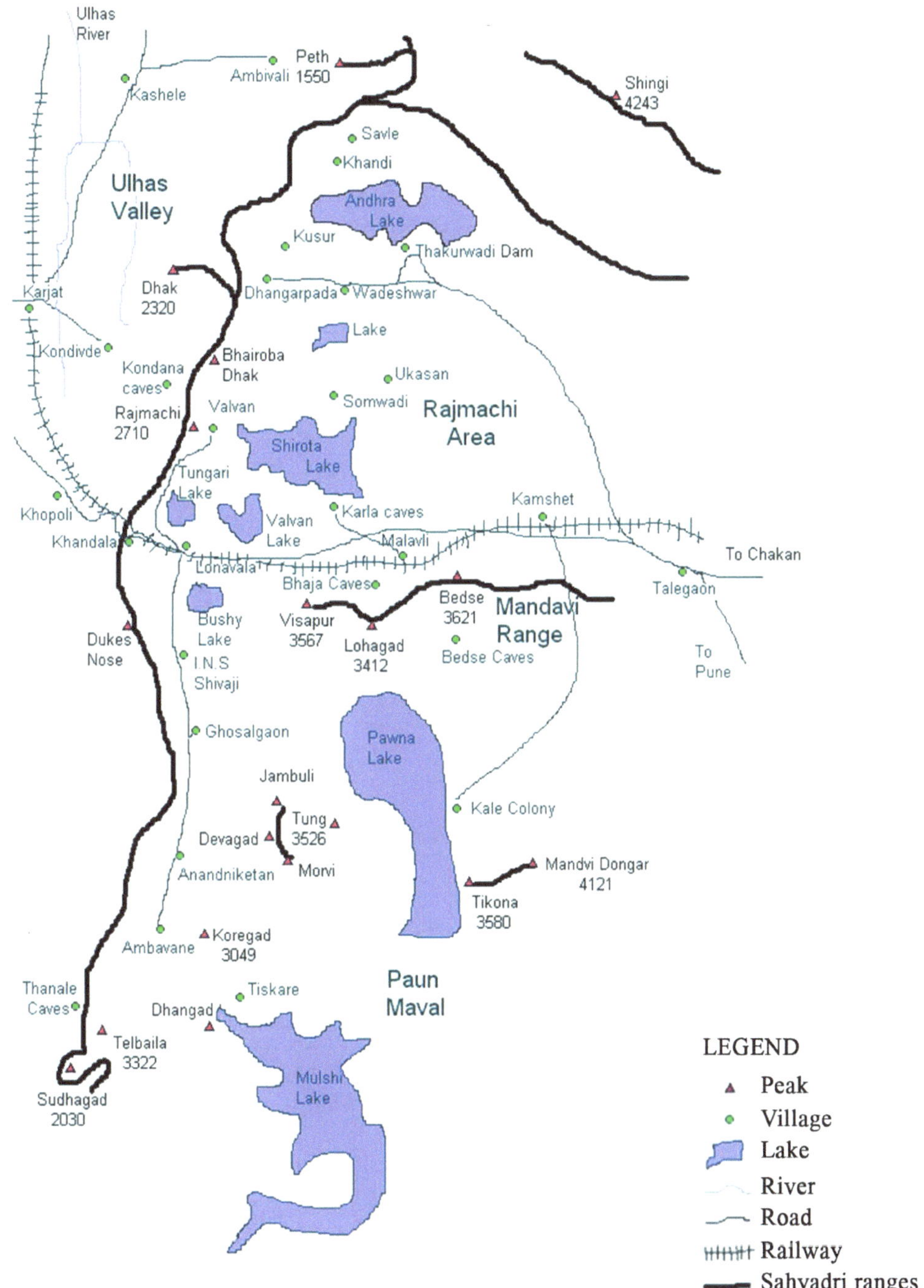

Figure 25 : Lonavala region

3.5 Tourist spots: Nature,Adventure and Heritage

Lonavala experienced a surge in tourism post-pandemic, particularly for weekend getaways, staycations, and destination weddings. The region's adventure tourism, supported by Maharashtra's Adventure Tourism Policy, has cemented its appeal for trekking, camping, paragliding and wellness retreats. As of 2021, Lonavala had around 2,060 hotel rooms, with independent hotels making up 75% of the market. Notably, the hospitality sector saw an increase in average rates by around 20% in 2022 compared to pre-pandemic levels, highlighting its resilience and expanding appeal (WTTC)(HVS).

Lonavala's tourism growth has not only enhanced its economy but also contributed to infrastructure development, making it a prime example of tourism-driven urban growth.

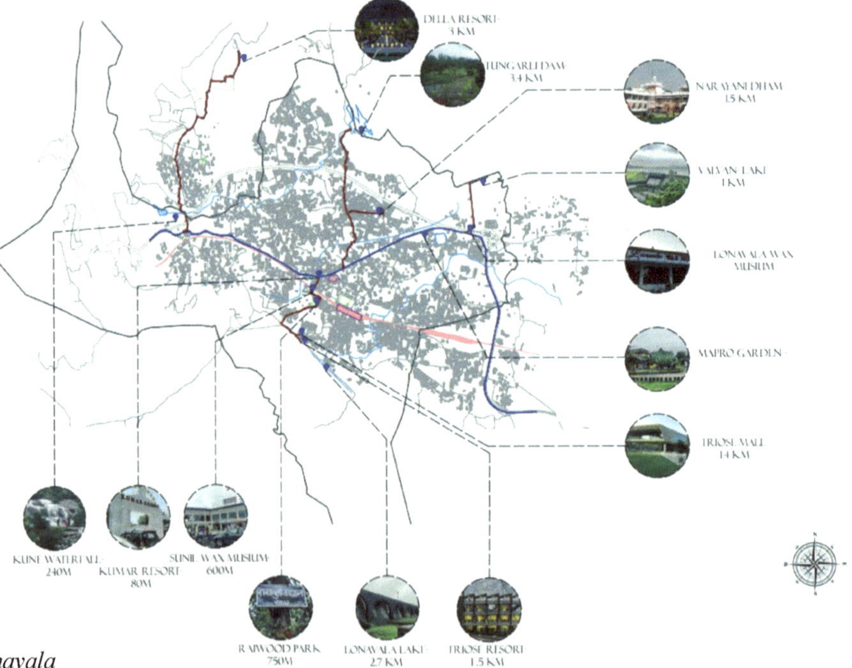

Figure 26 : Tourist spots in Lonavala

72

Lonavala attracts a distinct profile of tourists, primarily young professionals aged between 21 and 35. These visitors are typically part of middle management, entrepreneurs, or business owners with graduate or postgraduate qualifications. They predominantly fall within the socio-economic categories of SEC A1/A2 or B1/B2, with stated monthly incomes ranging from ₹10,000 to ₹35,000. Notably, most of these tourists drive to Lonavala, taking advantage of its proximity to Mumbai and Pune.

As a popular weekend getaway, Lonavala sees an influx of over 61,846 tourists every weekend. Additionally, approximately 1,600 tourists visit more than 5,000 privately owned bungalows and holiday homes each week. The town's 92 hotels and resorts, with a combined capacity of 2,446 rooms, maintain a 59% average occupancy rate throughout the year. A significant 78% of these visitors travel by car, underscoring Lonavala's status as a convenient road-trip destination.

The peak tourist season extends from June to September, coinciding with the monsoon, when the natural beauty of the region is at its best. Tourists typically stay for about three days, especially families, couples, and groups of friends. Corporate visitors and solo travelers tend to have shorter stays, averaging two days.

These statistics highlight Lonavala's steady tourism flow, supported by its strategic location, diverse accommodation options, and strong appeal as a leisure destination.

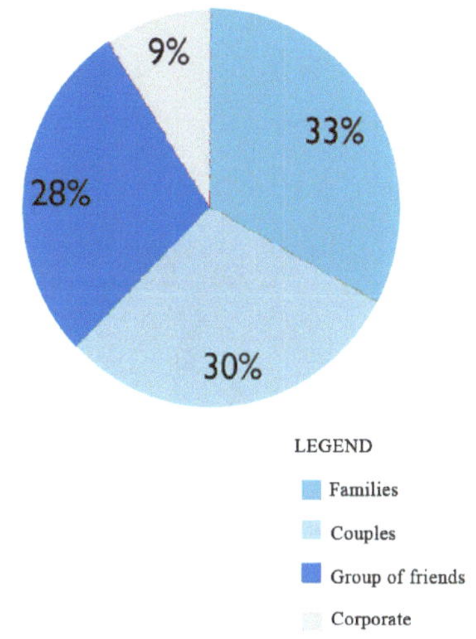

Figure 27 : Tourist flow

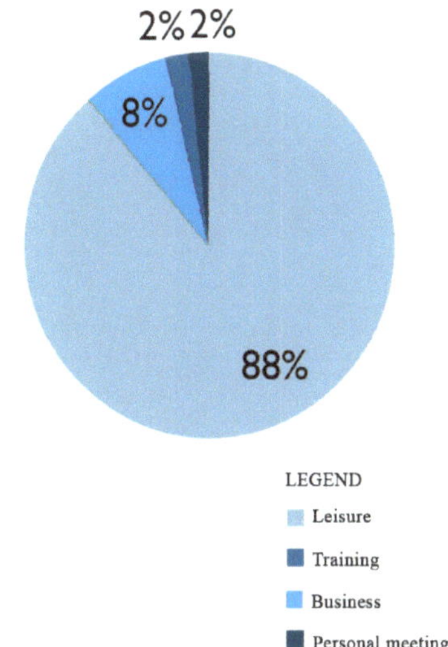

Figure 28: Purpose of visit

Image 37: Arial view of the built environment

3.6 Demographic Value

Lonavala, located in the Pune district of Maharashtra, operates as a Municipal Council and is administratively divided into 15 wards, with elections held every five years. According to the 2011 Census of India, the city had a population of 57,698, comprising 30,854 males and 26,844 females. Children aged 0-6 years account for 10.37% of the total population, or 5,983 individuals.

The female sex ratio in Lonavala stands at 870 females per 1,000 males, which is below the Maharashtra state average of 929. However, the child sex ratio in the 0-6 age group is 903, slightly higher than the state average of 894. Lonavala boasts a high literacy rate of 89.33%, surpassing the state average of 82.34%. Male literacy is recorded at 93.45%, while female literacy is at 84.57%, reflecting the city's emphasis on education.

As of 2023, the population of Lonavala is estimated to have grown to approximately 79,000, reflecting its steady demographic expansion.

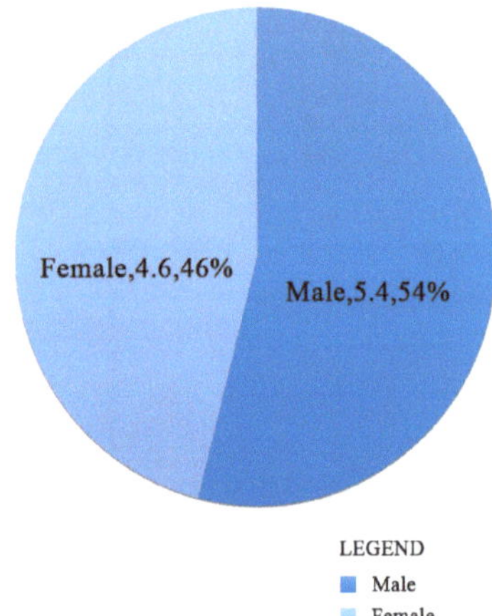

Figure 29:Percentage of male to female population

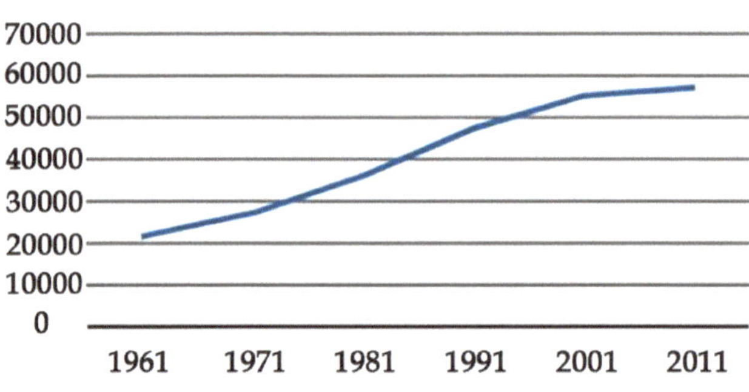

POPULATION (1961-2011)

Image 38 : Lane parallel to Shivaji Chowk showcasing naturally evolved mixed-use community

3.7 The Pulse of Commerce
The flow of economics

The markets and bazaars of Lonavala form an integral part of its socio-economic fabric, offering a blend of tradition and modernity. The town's commercial centers, such as the Lonavala Bazaar, are bustling hubs that serve not only the daily needs of locals but also cater to the substantial influx of tourists. These markets are particularly known for selling regional specialties, with chikki being the hallmark product that has garnered a nationwide reputation. Alongside this, items such as traditional Maharashtrian textiles, handicrafts, and a wide variety of local snacks draw in both residents and visitors. Locals in Lonavala exhibit a preference for traditional markets, where they shop for fresh produce, daily essentials, and artisanal goods. The shopping habits are largely influenced by the town's agrarian connections, with many farmers bringing their goods to the central markets, which are rich in fresh vegetables, fruits, and homemade products. During festive seasons, these markets become even more lively, with stalls offering unique handcrafted items, spices, and decorative materials that reflect the cultural richness of the region.

While traditional shopping remains dominant, Lonavala is gradually evolving with the presence of modern retail outlets and branded stores. This shift is largely a response to the increasing tourist footfall from nearby urban centers like Mumbai and Pune, who contribute significantly to the local economy. Tourists, in particular, frequent the markets to purchase souvenirs, artisanal crafts, and specialty foods, adding an important dimension to the town's commercial activities.

Key market areas such as the Tungarli Village Market and smaller markets near popular tourist spots, like Ryewood Park and Bhushi Dam, showcase a wide variety of products, from daily necessities to tourist-centric goods.

The diversity of offerings in these markets supports a unique economic ecosystem where both rural and urban consumption patterns intersect.

Despite the growing commercial and branded retail presence, Lonavala's traditional bazaars remain central to its identity, with their rich array of local goods and cultural significance maintaining their prominence in the town's economic and social landscape. These spaces reflect not only the evolving shopping habits but also the enduring importance of local craftsmanship and community-based commerce in Lonavala.

		Frequency of Visits						
Sr.no	Category	1-2 Times a	1 in 15-20 days	Once a month	1 in 2-4 months	1 in 6-8 months	Once a year	Less frequently
1	Eating Out	15%	16%	29%	24%	13%	3%	1%
2	Apparel Purchase	2%	2%	12%	37%	30%	12%	6%
3	Fruits and Vegetables	69%	13%	14%	3%	2%	1%	0%
4	Entertainment	6%	22%	15%	17%	8%	4%	15%
5	Daily Needs	71%	14%	13%	1%	1%	1%	0%
6	Stationary/ Toys/	2%	4%	16%	28%	27%	19%	4%
7	Consumer Durables	1%	3%	12%	9%	11%	39%	24%
8	Footwear	1%	1%	11%	28%	34%	16%	10%
9	Home Decor	1%	1%	3%	10%	24%	28%	33%
10	Food and Beverages	31%	24%	21%	11%	5%	5%	3%

Figure 30 :Shopping habits - Frequency of visits by locals

		Frequency of Visits						
Sr.no	Category	Less than 15	15-30 Min	31-60 min	1-2 hours	2-3 hours	3-4 hours	More than 4
1	Eating Out	5%	20%	34%	35%	6%	0%	0%
2	Apparel Purchase	8%	25%	37%	23%	6%	1%	0%
3	Fruits and Vegetables	32%	52%	12%	4%	1%	0%	0%
4	Entertainment	0%	8%	10%	14%	38%•	28%	2%
5	Daily Needs	22%	46%	26%	3%	2%	0%	1%
6	Stationary/ Toys/	15%	57%	19%	8%	1%	0%	0%
7	Consumer Durables	13%	36%	35%	14%	1%	1%	1%
8	Footwear	19%	44%	29%	5%	3%	1%	0%
9	Home Decor	14%	45%	32%	7%	2%	1%	0%
10	Food and Beverages	35%	49%	13%	4%	1%	0%	0%

Figure 31 :Shopping habits - Time spend

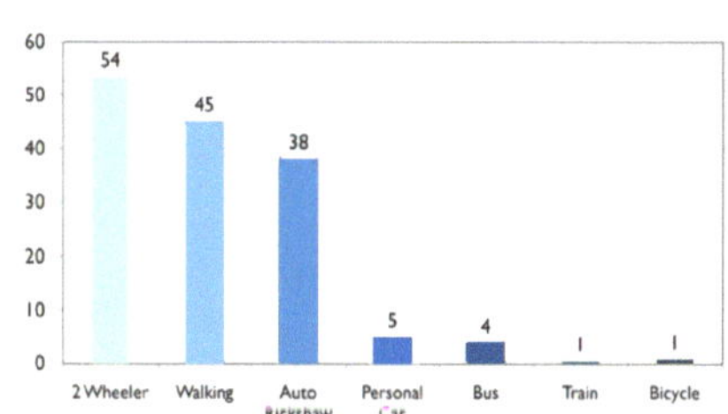

Figure 32 :Mode of transport used by locals to visit the market

3.8 Nodal Anchors of Economics

The node situated between Kumar Resort and Fariyas Resort, along the Mumbai-Pune Highway (NH-48), serves as a critical junction within Lonavala's urban landscape. This area is characterized by its dual role as a transit hub and a destination for tourists, heavily influenced by the region's scenic beauty and climatic conditions.

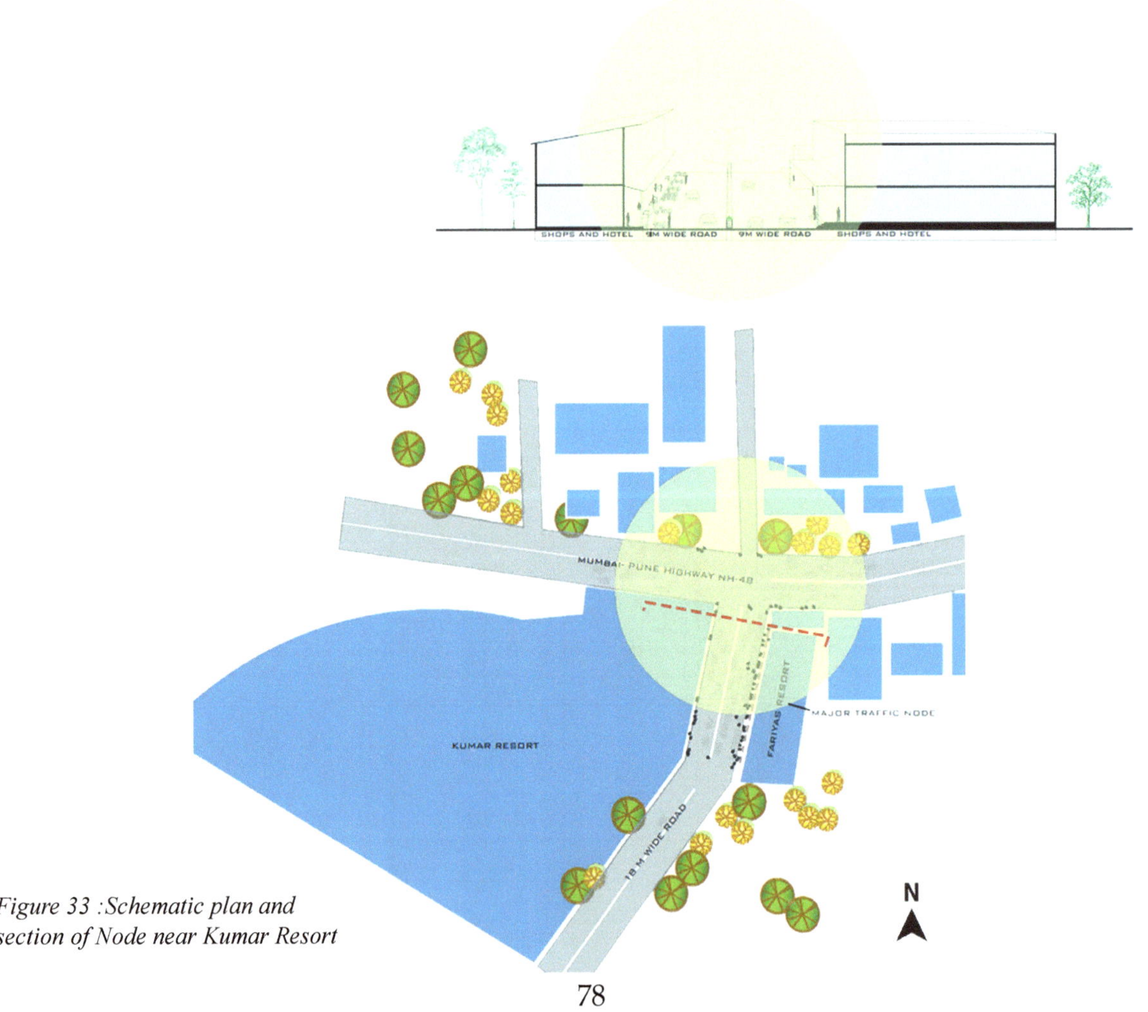

Figure 33 :Schematic plan and section of Node near Kumar Resort

Image 39: Main bazaar road leading towards Mumbai-Pune highway

Several factors affect the existing urban context of this node. Firstly, the proximity to major transportation arteries facilitates access, making it a focal point for commercial activities and hospitality services. The high volume of traffic along NH-48 enhances visibility for businesses, yet it also contributes to congestion and environmental concerns.

Secondly, the presence of established resorts and recreational facilities underscores the node's importance in the tourism economy, attracting both weekend visitors and longer-term guests.

Lastly, land use dynamics play a vital role in shaping the urban fabric of this node. The presence of high-density resorts juxtaposed with low-density residential areas creates a diverse urban experience.

The area between the Lonavala Municipal Council (LMC) Market and the Municipal Market, including the prominent Shivaji Putla Square, is a vibrant hub that reflects the town's socio-economic dynamics. This node serves as a critical intersection for local commerce, community interactions, and cultural activities.

The LMC Market is a key center for daily needs, offering a variety of goods ranging from fresh produce to household items, catering primarily to the local population. Conversely, the Municipal Market provides a diverse array of retail options, attracting both residents and tourists. The juxtaposition of these markets fosters a lively atmosphere, encouraging foot traffic and enhancing local business opportunities.

Figure 34 :Schematic plan and section of Node near Shivaji chowk

Shivaji Putla Square, adorned with a statue of the revered Maratha king, serves as a cultural landmark and gathering point, hosting events and celebrations that bolster community engagement. However, the area faces challenges such as traffic congestion and inadequate pedestrian infrastructure, which can hinder accessibility and safety.

Image 40: Shivaji chowk

Image 41: Arial view of the Indrayani river

4. TOWARDS SUSTAINABLE LONAVALA

Lonavala, a scenic eco-sensitive location in Maharashtra, has evolved due to its assorted ecological, physical and economical dimension, pioneer in shaping the region's lifestyle, landscape and ecosystem. The investigation encompassed comprehensive ecological surveys to understand the biodiversity, ecosystem dynamics, and environmental challenges, faced by the region. A detailed study of each aspect has provided valuable insights into the region's evolution. The socio-economic aspects influencing the local primary business, of tourism, agriculture and food processing, also impacts the eco-sensitive environment of the region. Analytical classification of these inferences under strengths, weakness. Opportunity and threats help evaluate and plan the future for Lonavala.

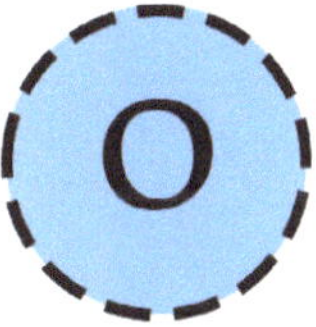

STRENGTH	WEAKNESS	OPPORTUNITY	THREATS
- Rich in flora and fuana - It has potential for ecotorism - Flourishing hospitality industry	- Environmental degradation - Lack of free space in compact urban fabric - Deforestration - Seasonal dependency on tourism, leading to fluctuations in economic activity	- Can promote balanced ecosystem. - To promote local craft and hand crafts - Boosting socio economic status of community	- Excessive tourist lead to overcrowding - Natural disasters like landslides and floods.

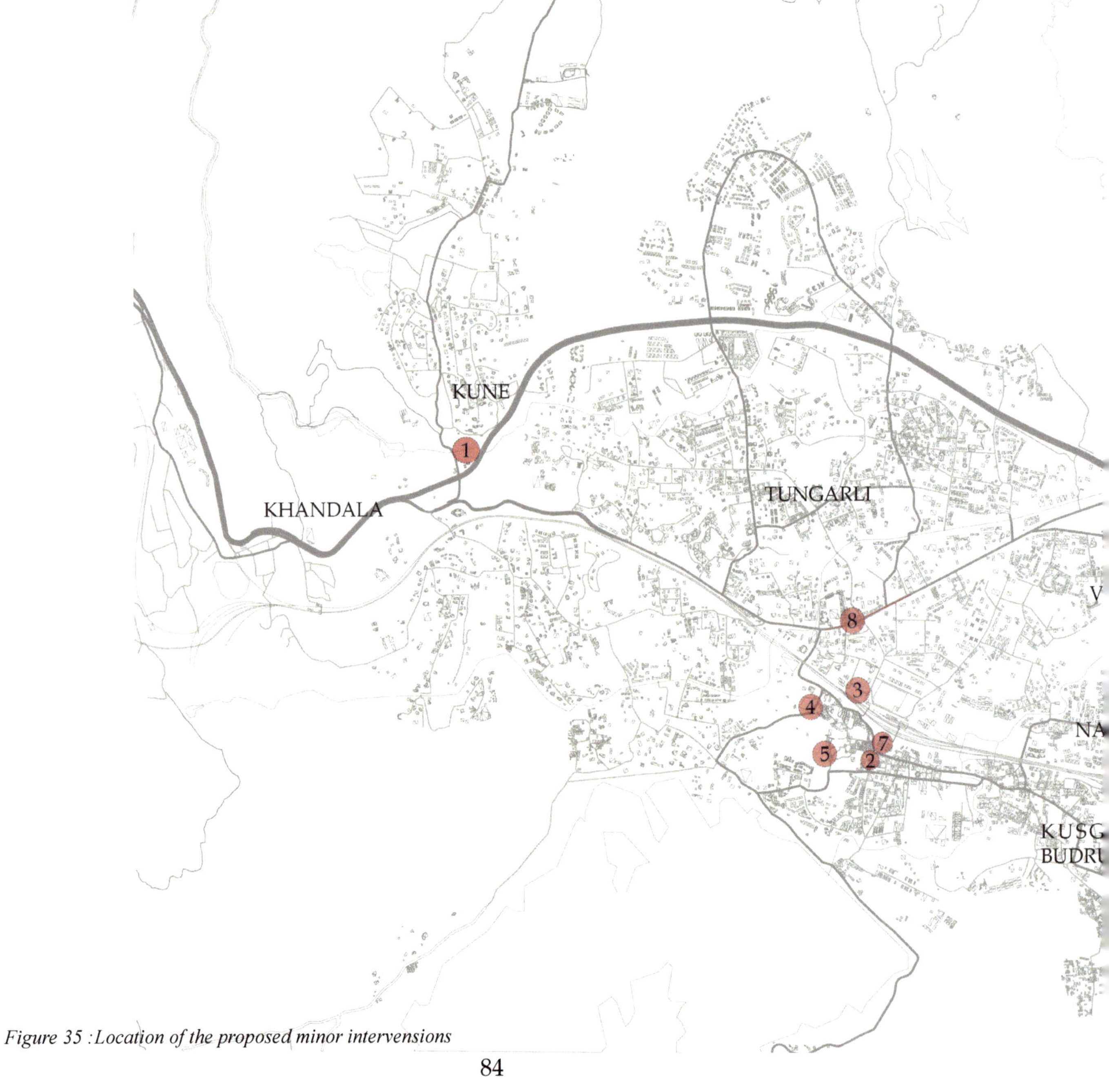

Figure 35 : Location of the proposed minor intervensions

4.1 Minor Intervensions towards Sustainable Living

The study delves into various aspects of design, focusing on how targeted minor interventions can enhance public spaces, traffic management while maintaining ecological balance. It investigates the strength of these measures like boosting tourism, improving traffic flow and providing public amenities and the challenges they face, particularly in terms of urban governance and seasonal population surges. By examining key nodes like Shivaji Maharaj Chowk, the railway station, Raywood Park etc. it explores how urban improvements in these areas can have far-reaching impacts on the city's overall functionality and sustainability.

Although there is a need for alignment with Maharashtra's Regional and Town Planning (MRTP) Act and effective collaboration between the Lonavala Municipal Council (LMC) and local stakeholders to ensure successful implementation.

1. Bird Nesting Tower , Kune.

2. Street Re-Structuring, Shivaji Maharaj Chowk Lonavala.

3. Space For Car Parking, Railway Station.

4. Garden Libarary, Thombrewadi.

5. Enhancing Area/Open Inside The City,Raywood park

6. Lake Front Development, Valvan

7. Mobile Kiosk for street vendors , Shivaji Maharaj Chowk Lonavala

8 User Friendly Streets

BIRD NESTING TOWER

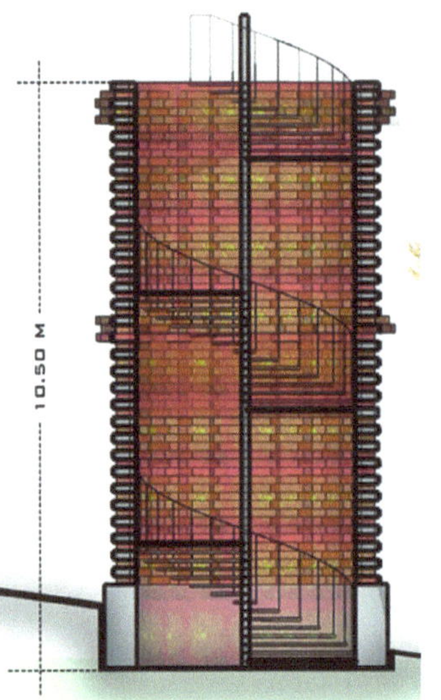

Figure 36: Schematic repersentations for the bird nesting tower

The primary function of nests tower is to provide a suitable location for birds to lay their eggs and /or raise their offspring. The design and function of bird's nests is far more sophisticated than previously realized. Nevertheless,there are still several locations in lonavala where these towers can be built, that are likely to be fruitful for these birds. The breeding season for birds is different across all the species. This bird nesting tower can help cater to the varied seasonal breeding throughout the year .

Material used-

The material used for the bird nesting tower is laterite stone. It is used since it is a natural Material which has cooling properties. This material can be procured from karjat which is 70kms from lonavala.

The high water absorption capacity of the material suggests adoption suitable damp protection methods to improve performance of laterite masonry structures. The stone hardens and gains strength as time progresses. Environment friendly as they do not emit CO_2 and greenhouse gases

STREET RE-STRUCTURING

The intervention proposed is street re-construction, which explains the street which are of existing 12m to be used at its fullest with dedicated greenary and pedestrian walkways acting as a buffer between main vehicular road and existing shops.

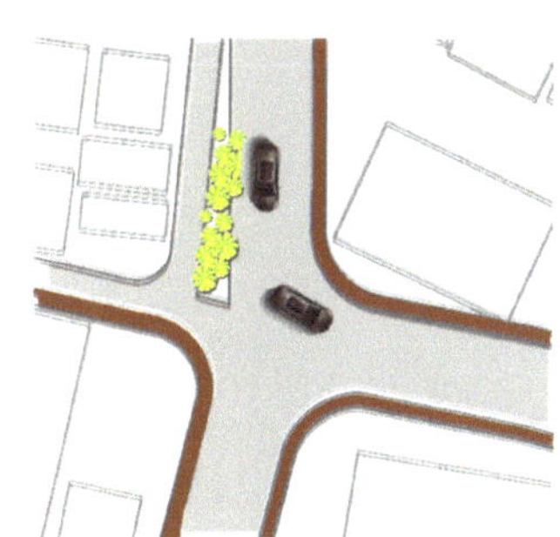

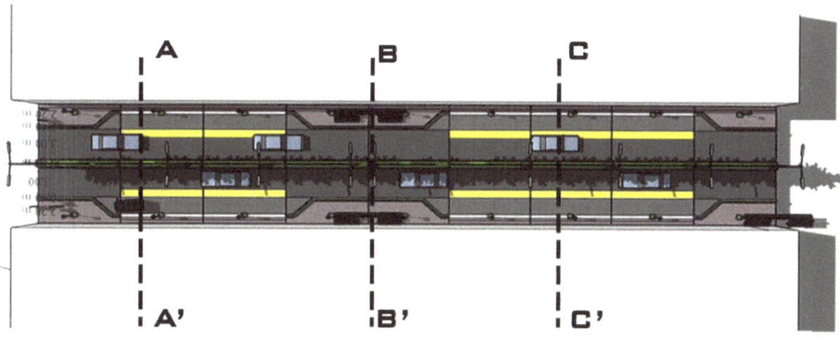

SECTION BB'

Figure 37: Schematic repersentations for street re-structuring

Street view

Sector 1

Sector 2

SPACE FOR CAR PARKING

The street is an important east-west urban connectore, with ground floor retail frontable,along the corridor,which generates high pedestrian and vechile volumes everyday. The common issues observed on this road included lack of walkable footpaths,vehicular congestion, bottlenecks, multiple layers of street side parking and poor pedestrain crossing infrastructure

Narrow lanes reduce speed and minimize ceashes on city streets by way of reducing the right of way and making drivers way of traffic and adjacent users.The additional space for pedestrain space, cycle facilities, or green infrastructure

This area is accessible by primary and secondary roads.Primary road has width of 9m and secondary road has width of 6m

The site is near lonavala railway station. the site area is about 450 sq.m

Thousands from Pune, Mumbai and adjoining places visit lonavala during wekends in the monsoom. Close to 5000 vehicles arrives at lonavala every wekend while 1000 vehicles visit the hill station on week days. Tourists are forced to park along the roads of the town which causes traffic jams

The authorities have made available some private parking plots but they are not enough. These parking plots are located close to the town and away from the scenic points. People do not park their vehicles here.It was imperative for LMC to make available plots at tourists points where people can park safely

The car parking area is small land.No one is there for security of cars and anyone can come and park their vehicle there and cars are also not park properly in monsoon.The car parking area is muddy area because of rain water

Proposal of proper car parking near the railway station design proper car parking area with dividers also

The proposed design of lonavala main road included the following principles

Maintaining contiguous vehicular trahhic lanes two lanes on each side creating multi-utility parking zone all along the street introducing protected pedestrian crossings

Introducing "Placemaking" to manage the footpaths

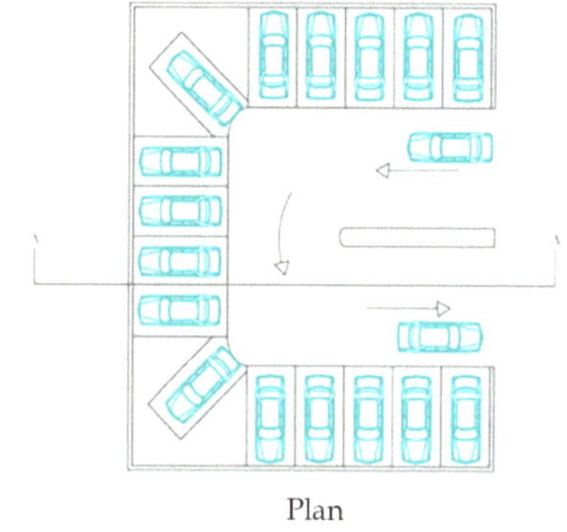

Plan

Section AA

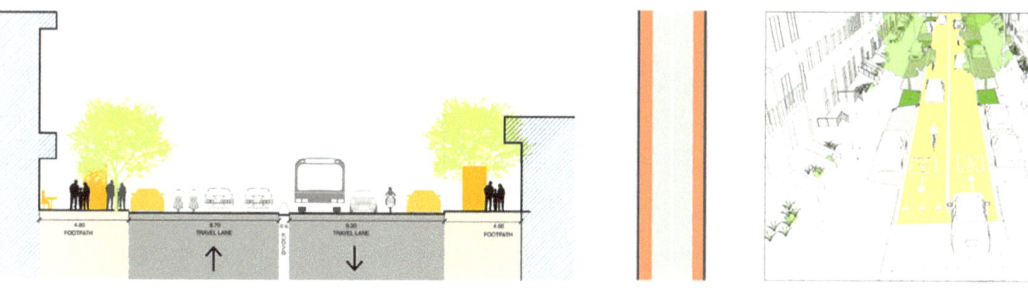

Figure 38: Schematic repersentations for space for car parking

GARDEN LIBRARY

WHAT IS GARDEN LIBRARY?

Library garden aims to reclaim the forgotten love of paperbacks and separate readers from their phones, tablets, electronic readers, and wherever else they consume written content these days.

BASIC IDEAS:

• We all know that Lonavala is a tourist Destination. People often visit lonavala to spend some quality time and relax and to take a break from their busy schedule

• But due to this the local people have to deal with all the congestion

• While talking about the local people, youth /the students are a major part of that. If we look at the urban fabric of lonavala we can see that the students of lonavala don't have a peaceful environment to study other than their school, coaching centers and home.

• Taking that into consideration we would love to introduce our intervention that would help encourage healthy learning options by creating student friendly open-air garden libraries/ outdoor study café.

SITE:

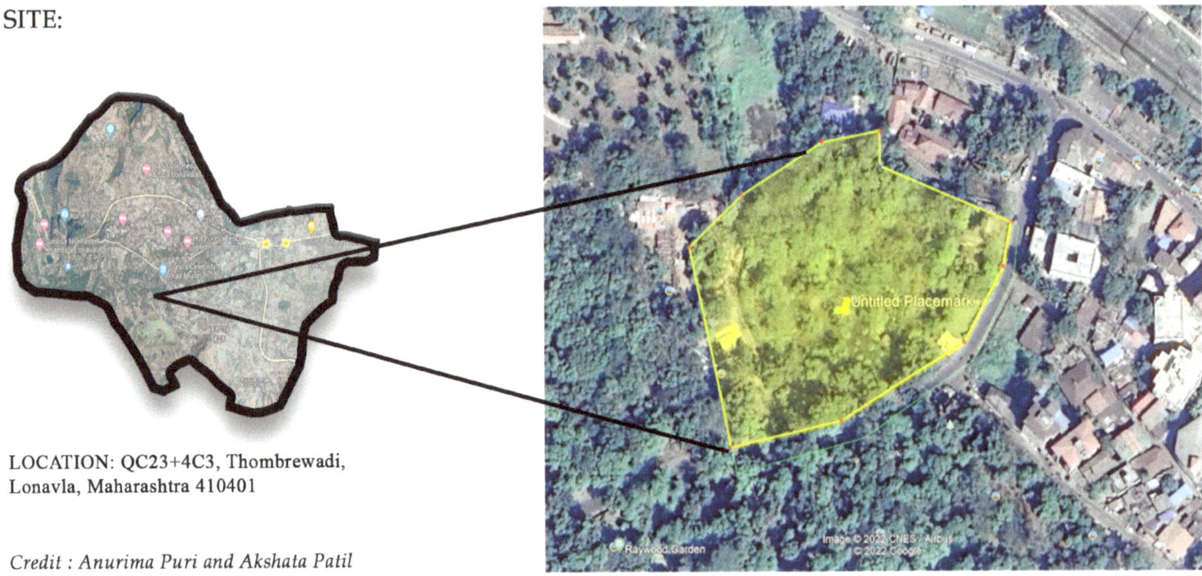

LOCATION: QC23+4C3, Thombrewadi, Lonavla, Maharashtra 410401

Credit : Anurima Puri and Akshata Patil

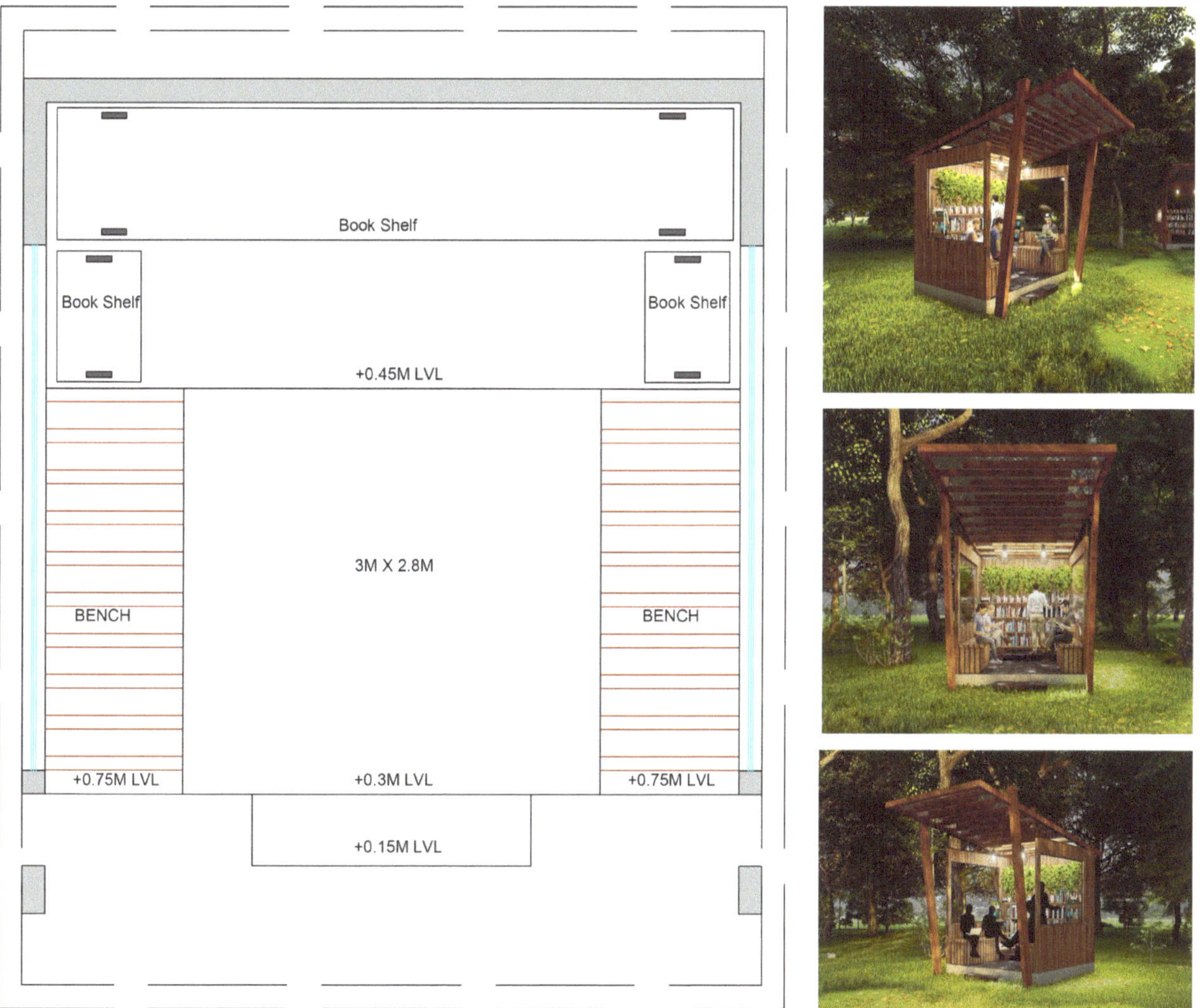

SEMI OPEN LIBRARY PLAN

Figure 39: Schematic repersentations for garden library

ENHANCING GREEN / OPEN INSIDE THE CITY, RAYWOOD PARK

In terms of environmental objective, it can help to reduce air & noise, as there would be a reduction in the no. of vehicles parking on the street side.

All trees around the site are maintained. We designed for more space on the street not only fpr pedestrians but also for planning of additional planning areas, improving street furniture and landscapings.

WHAT TYPES OF GREEN SPACES CAN A CITY HAVE?

1. Urban Farm.
2. Green Houses.
3. Vertical Farming.
4. Beehives and Highways.
5. Wildlife Corridors.
6. Integrated Habitat reations.
7. Flood Residence.
8. Water Storage.
9. Sustainable Urban Drainage.
10. Bioremediation.
11. Green wall-top down.
12. Green wall.
13. Modular Plant WAlls.
14. Seeded Living Walls.
15. Moss Walls.
16. Tree Facade.
17. Bioreactive Facade.
18. Green Roffs.
19. Wildlife Roofs.
20. Wet Roffs.
21. Urban Vegetation.
22. City Gardens.
23. Photovoltaic Roofs.
24. Wind Turbines and Micro-Generation.

ENHANCING GREEN / OPEN INSIDE THE CITY (RAYWOOD PARK, LONAVALA)

Figure 40: Schematic repersentations for enhancing open space in city

Credit: Prayag Khandalkar and Chetan Joshi.

VALVAN LAKEFRONT DEVELOPMENT

Lakefront design refers to the careful planning and development of spaces adjacent to a lake. It involves creating aesthetically pleasing and functional environments that take advantage of the natural beauty of the waterfront while providing various recreational and leisure opportunities for residents and visitors alike. Lakefront designinclude elements such as walkways, parks, gardens, seating areas. Additionally, the integration of environmentally sustainable practices and preservation of the lake's ecosystem are essential considerations in such designs. Overall, lakefront design seeks to create inviting and harmonious spaces that allow people to connect with nature and enjoy the serenity of the waterfront setting.

SITE:
Area: 3.21 Acre

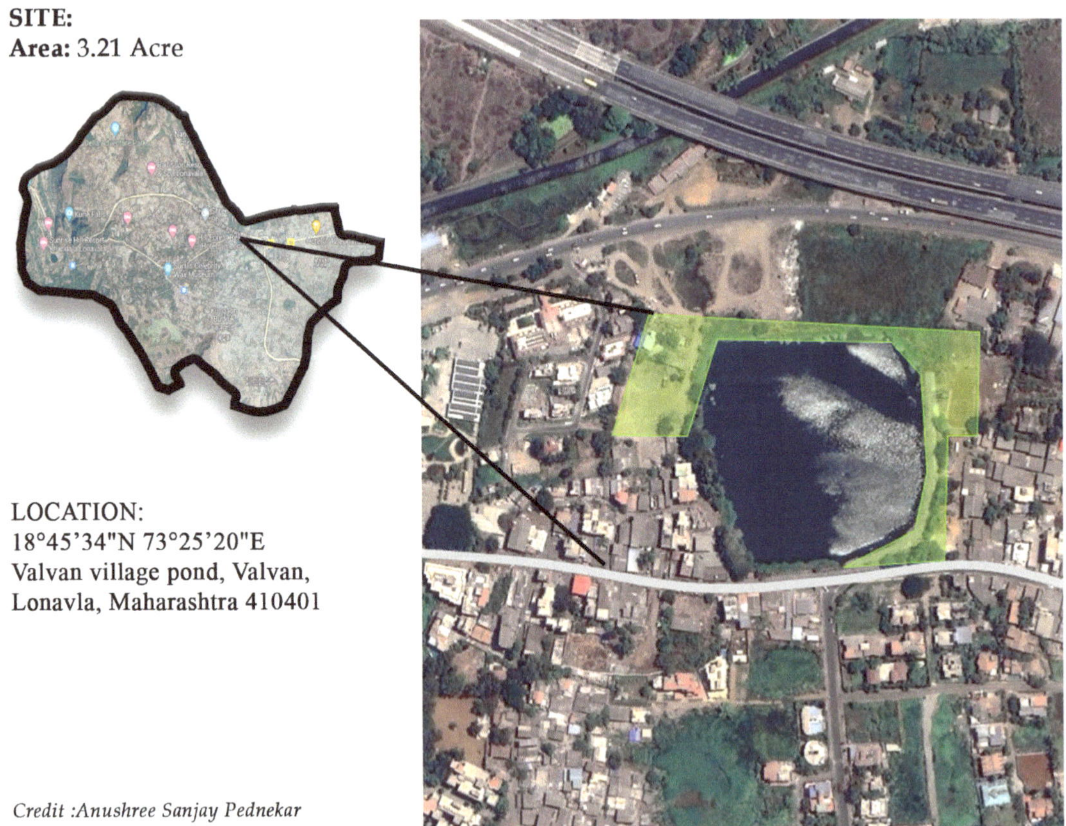

LOCATION:
18°45'34"N 73°25'20"E
Valvan village pond, Valvan,
Lonavla, Maharashtra 410401

Credit :Anushree Sanjay Pednekar

Image 42: Valvan lake

CONCEPT

The motive is to design a Green park which not only act as green space in Urban city but also create community driven space.

LINES

To aid the pedestrian flow through the Garden, the paths have been realigned, creating Route of movement interconnecting different spaces and edged with integrated seating to accommodate the large number of visitors the Garden receive

POINTS

More planting beds have been introduced, all designed to attract different species of Birds and Butterflies.
Tree avenue incorporating rock garden and sensory garden, creates southing sound and experience along the movement.

SURFACES

The design layout creates number of different spaces; some quite and intimate, some busy and open, all acknowledge the needs of resident & tourists.

A Parking	**D** Viewing Bridge	**G** Play Ground	**J** Sensory Garden with Musical Sculptures	**L** Open Lawn Area for Meditation/Yoga
B Entrance Gate	**E** Seating Area	**H** Open Gym Area	**K** Amphitheater	**M** Bird-friendly Garden
C Green Wall	**F** Kids' Play Area	**I** Rock Garden		

Figure 41: Schematic repersentations for Valvan lake front development

G
H
I
J
K
L
M
D
F
E
C
B
A
Valvan Village Pond
PUBLIC TOILET
9m wide Road
N
E
S
W

CONTROLLING SPILL OFF OF STREET :
Vendors by Providing Mobile Kiosk In Market
Location : Shivaji Maharaj Chowk

The site is near Shivaji Maharaj Putla where a local everyday market it situated. The Lonavala Municipal corporation building, which is just behind the Statue, Is accessible by a 15 meter road. However, the road is encroached by hawkers and street shops which are temporarily made tents. This increases the traffic congestion in the area despite of the sufficient road width.

Design intervention: The intervention focuses on segregation of pedestrian, hawkers, shops and vehicles on the street to avoid traffic congestion and
increase safety for pedestrians. Along with this, there is incorporation of a small public plaza as an add on to the street market.

Figure 42: Schematic repersentations for mobile kiosk in market

PROTOTYPE:1
SIZE : 3.00 X 5.00 M

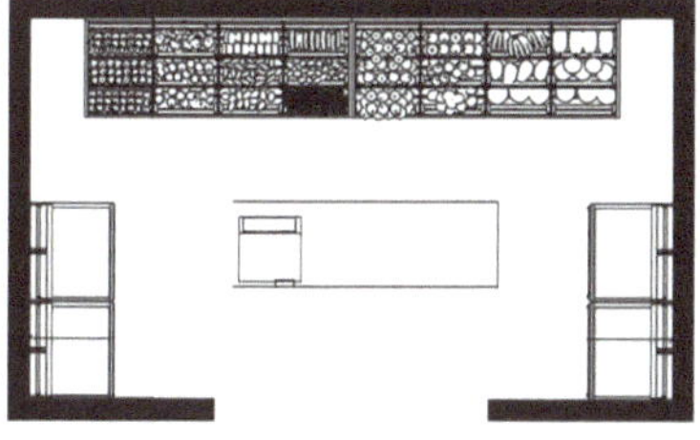

PLAN

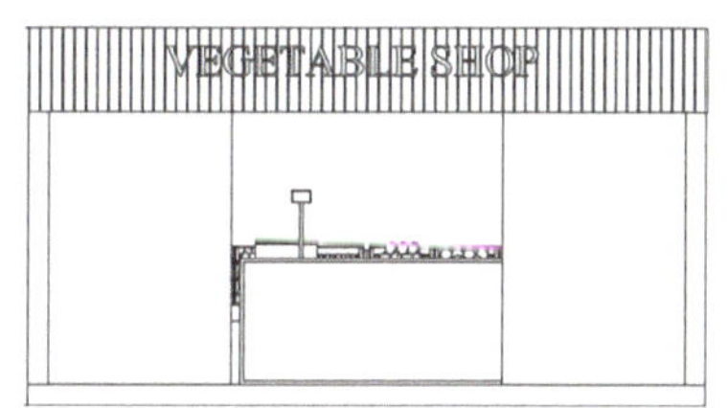

FRONT ELEVATION

Intervention Application : The proposed Intervention can be applied at Jaichand chowk which is near the Lonavala railway station. The site has similar condition where temporary shops are built causing traffic congestion on the road.

Materials :- The material used for kiosks is corten Steel shipping containers.

Credit : Heena Sharma, Aoudhut Mhatre , Rashmi Deshpande

MAKING STREETS MORE EFFICIENT AND USER FRIENDLY

The intervention here is to increase ease in mobility of traffic by bifurcation of road networks and the users also by providing multifacted activities in and around the streets so as to make them a quality public space.

Figure 43: Schematic reperesentations for making streets more efficient and user friendly

Crebit :Aarya Deshpande and Anuj Kumbhar.

Main road traffic analysis
With the increasing tourism population,the street scaping is not considered accordingly. The users do not have enough street space and stable traffic mobility

A pretty straigntforward motion that leaves little to no"
Multi-interpretation .bear in mind. however that " countries with proper public transport modules " should be immediately translated as "First-world countries".
As those countries are only ones who "Are likely to have proper public transport modules.

| 8AM TO 12PM | 12PM TO 6PM | 6PM TO 10PM |

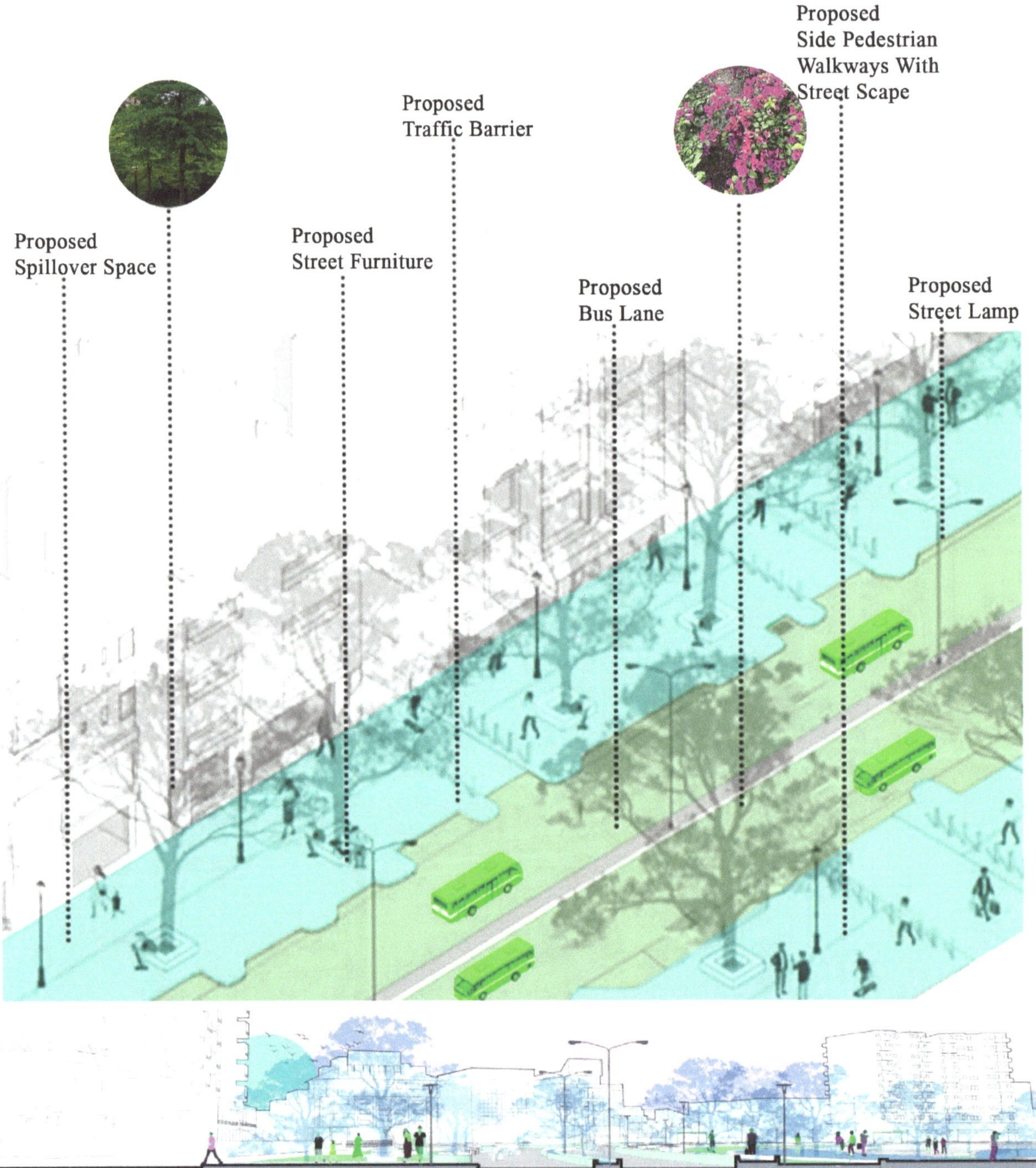

Proposed
Spillover Space
Proposed
Street Furniture
Proposed
Traffic Barrier
Proposed
Bus Lane
Proposed
Side Pedestrian
Walkways With
Street Scape
Proposed
Street Lamp

Objective

		Footpath		Shared space

Reduce parking

	Service Lane		No Service Lane

Introducing Neighbourhood level public Trasport

Divided Carriageway	Undivided Carriageway	No Carriageway

	Footpath		Shared Space

Incorporating Building Frontages to develop Predestrian realm

	Service Lane		No Service Lane

Divided Carriageway	Undivided Carriageway	No Carriageway

Proposed street with bus rapid transit

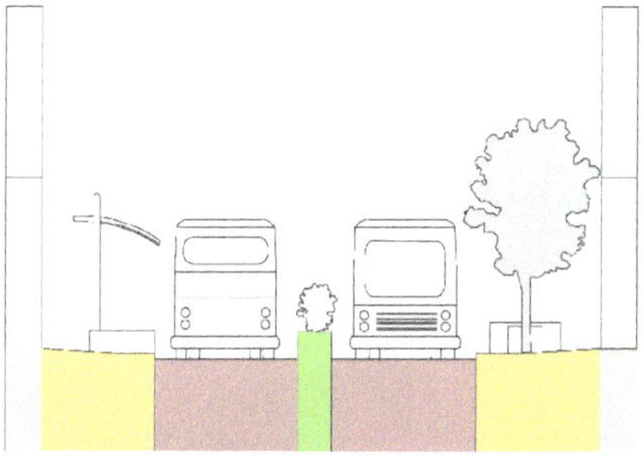

Pedestrian mobility and access

Parking and property access

Public vehicle mobility

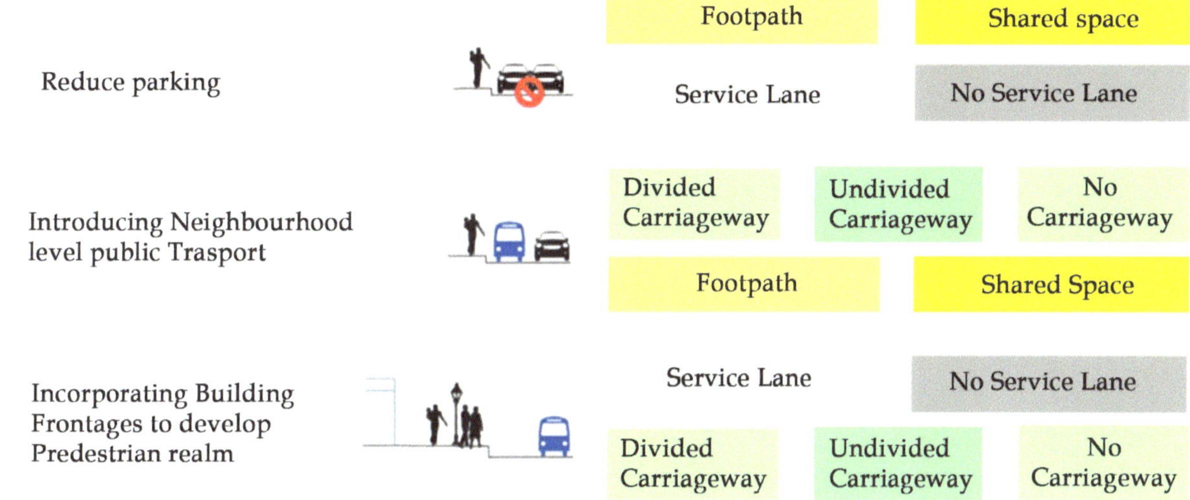

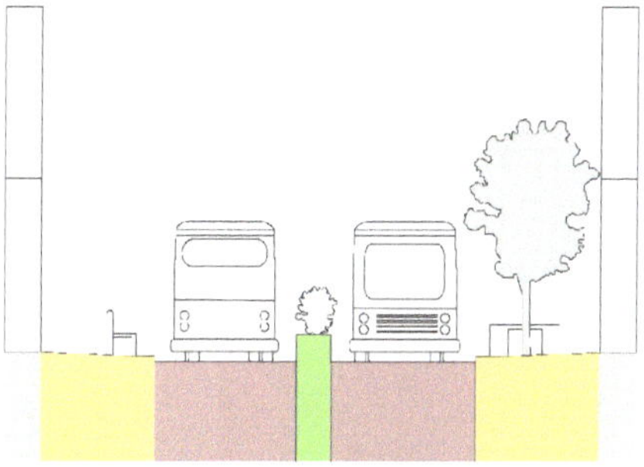

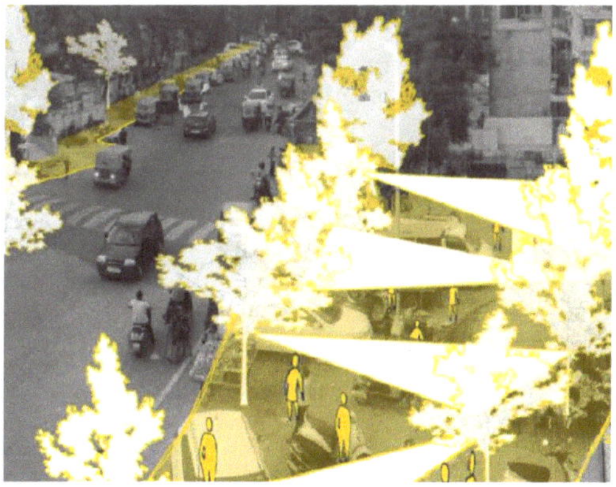

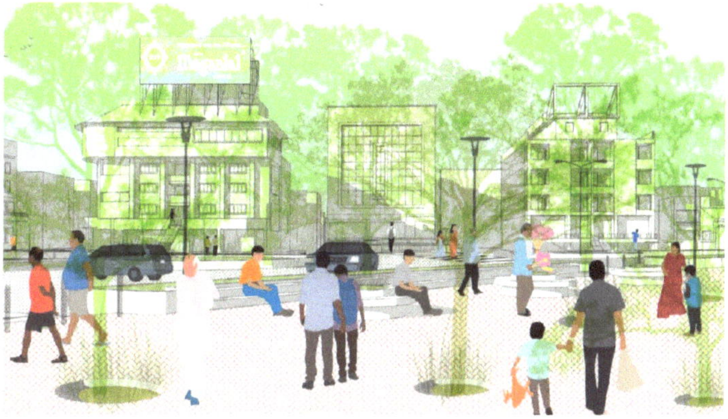

- Bus terminal design at Purandar ground.
- Bus terminal design at Valvan pond.
- Bus terminal design at hospital.

Figure 44 : Location of the proposed minor intervensions

4.2 Shaping The Cityscape: Significant Interventions

The synthesis of ecological, physical and economical aspects, contributes to a holistic understanding of a thoughtful way forward - towards sustainable futuristic development of Lonavala. Public transport holds center stage in the urban transport agenda. A well-functioning and sustainable city cannot be achieved without strengthening its public transport system. The National Urban Transport Policy (NUTP, 2014) focuses on public transportation by encouraging greater use of public transport, establishment of quality focused multi-modal public transport systems that are well integrated with essential services and provides seamless travel across modes and nodes.

The study reveals the major mode of public transport in the city of Lonavala being a Bus terminal and railway station with the former being predominant. A bus terminal is a point at the start/end of a bus route, where the vehicles stop, reverse and wait, before departing on the return journey, also serving as a station for passengers to board and alight. Currently redundant in it's services and facilities, the bus terminal located in the heart of Lonavala city exhibits tremendous potential for a community-oriented sustainable development.

Understanding the city at depth helped in identifying three possible locations for development of an integrated bus terminal with thoughtfully curated community engagement services which would survive symbiotically with the newly proposed bus terminals having state of the art facilities. The three prominent locations at Purandar ground, near Valvan lake and near hospital located at the old bus terminal, were identified for modal and nodal development. Integrated services like sports arena, hospital and evening and weekly markets (bazaars), were respectively amalgamated with bus terminal facility at Valvan lake, old bus terminal and Purandar ground sites. The greatest challenge involved in the proposals is to integrate the transport node with formal spaces and recreational community engagement activities.

Integrated bus terminal with co-working spaces
(Purandar ground)

CONCEPT

Mass And Void

Convex or concave forms are used in Architecture . As concave forms offer the appearance of void and convex ones give the impression of mass

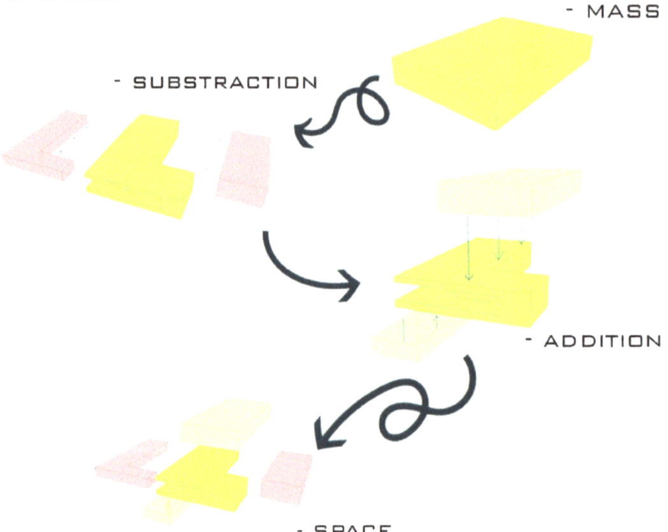

The bus terminal was created with the idea that buses serve as an essential link between two locations, fusing modern design with historic architecture through the use of domes and circular windows.

Credit: Aoudhut Mhatre

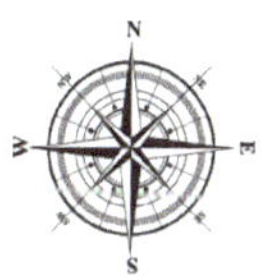

B
C
10 M WIDE ROAD
ENTRY
12 M WIDE ROAD
MARKET
A
A'
BUS STAND
CAR EXIT
CAR ENTRANCE
AUTO ENTRANCE
AUTO EXIT
10 M WIDE ROAD
WORKSHOP
12 M WIDE ROAD
EXIT
B'
C'
N
S
E
W

B
C
A
A'
90 CAR PARKING
80 BIKE PARKING
B'
C'

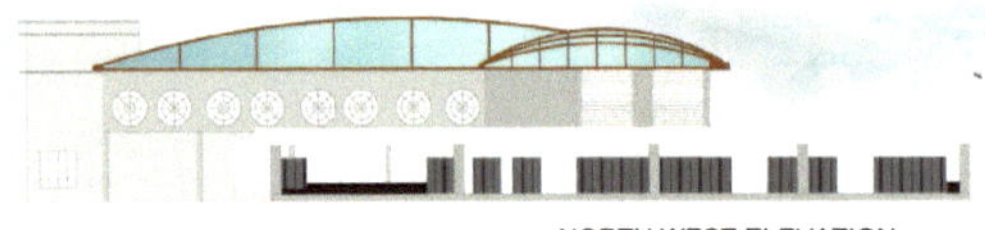

NORTH WEST ELEVATION

SOUTH EAST ELEVATION

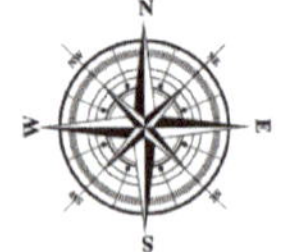

B
C
ENTRY
12 M WIDE ROAD
A
A'
WORKSHOP
MARKET PARKING
CAR EXIT
6 M WIDE ROAD
6 M WIDE ROAD
CAR ENTRANCE
AUTO ENTRANCE
AUTO EXIT
12 M WIDE ROAD
EXIT
B'
C'

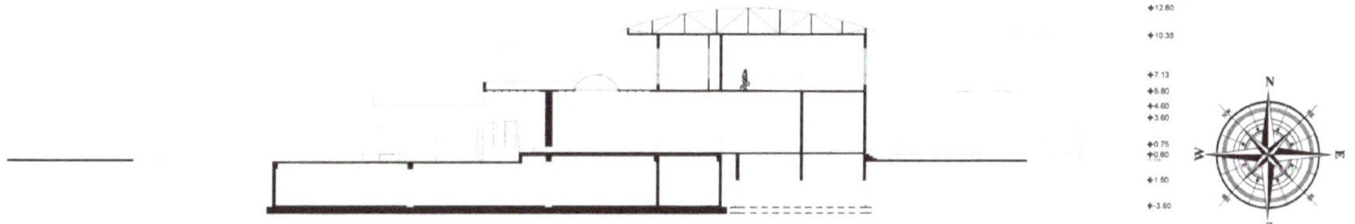

Section A-A'

FIRST FLOOR PLAN
B
C
5
ENTRY
12 M WIDE ROAD
A
A'
18 M WIDE ROAD
4
6
1
2
12 M WIDE ROAD
EXIT
B'
C'
Section B-B'
N
W
E
S
Section C-C'

3D View Of Bus Terminal

Integrated bus terminal with multi-purpose public arena (Purandar ground)

The practise of architecture is essentially social since it integrates the demands of the user, the larger community, and the built environment to create an ecosystem. By proposing bus terminal here the gap between urban and rural area what is due to the concrete jungle.

The project's core goal is to create a generative integrated bus terminal that emphasises a friendly and healthy social environment and gives the underrepresented a boost in the growth of the Lonavala. every place is unique and special so the speciality of the lonavala is tourism, so keeping tradition in this modern era so my design based upon the concept,
'MODERNITY IN TRADITION'

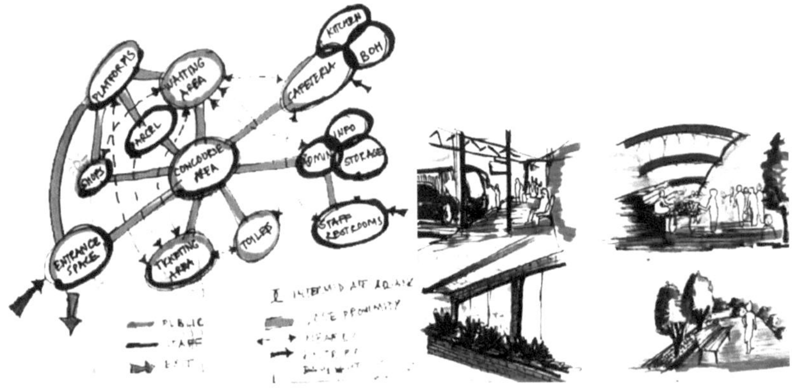

Conceptual zoning   Conceptual sketches

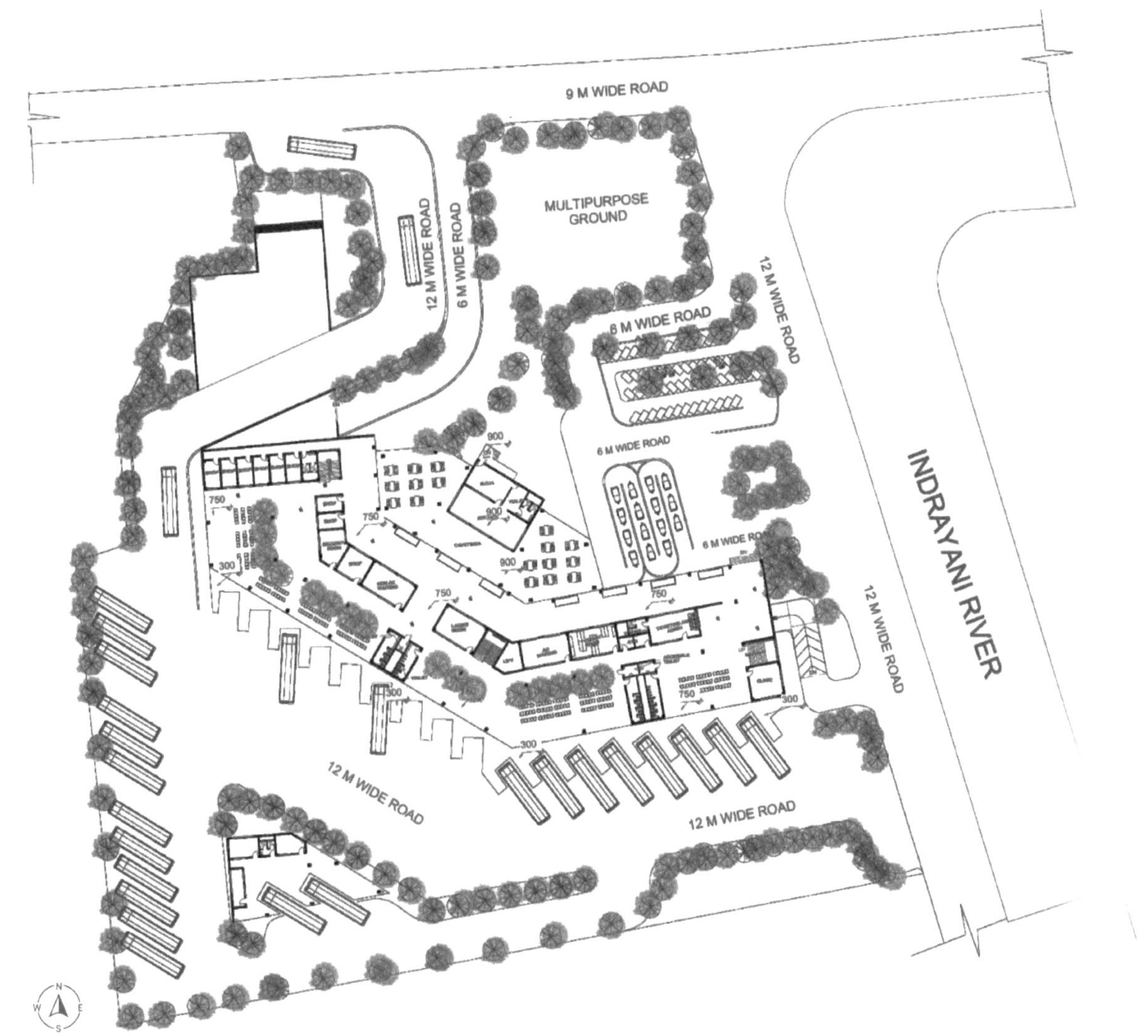

Ground floor plan

Exploded view

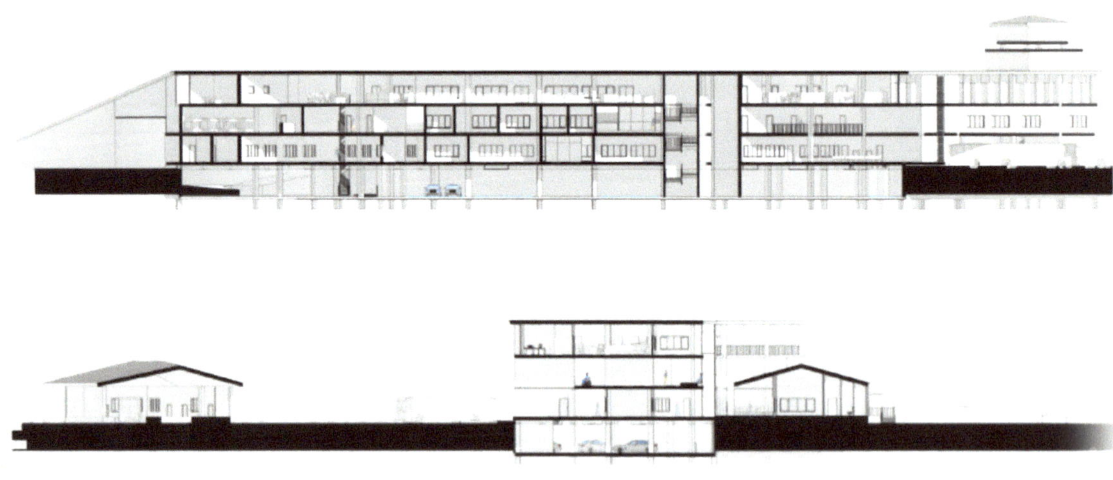

Sections

Bus terminal- View 1

Bus terminal-View 2

Integrated bus terminal with sports complex
(Valvan lake)

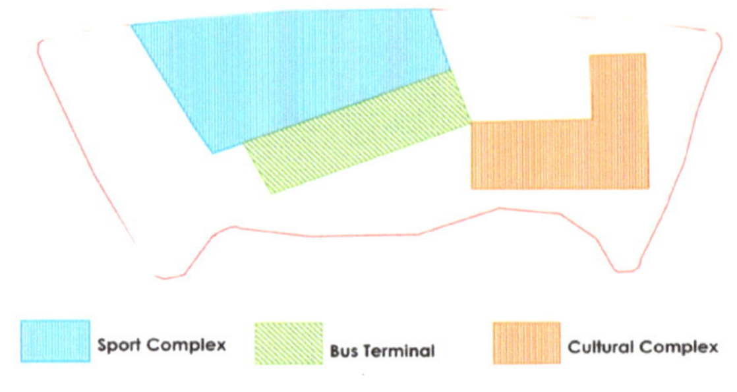

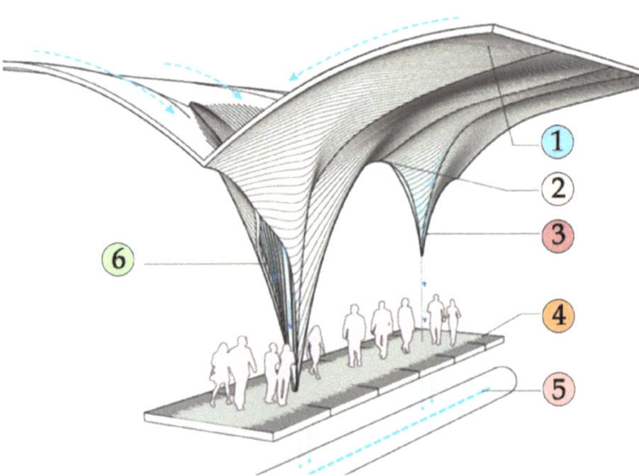

CONCEPT: FUNCTION WITH BOLD SCULPTURAL EXPRESSIONISM

• Free flowing
• Contrast
• Character
• Vibrancy and Excitement
• Connecting Journey
• The main design approach to create the design spaces where people intervenes planning spaces where people can enjoy short time, socialize and culture can overlap, gather around for function forming the interaction and interference with water

1. Perforated Edge: Perforations through the concrete at the periphery of the canopy allow light and water to pass.

2. Arches: High arches maintain street sight lines and OWS clearances.

3. Stalactite: Water leaves the canopy via a rain chain hung from canopy low points.

4. Rumble Strip: Cast into precast units.

5. Drain Pipe: Water directed from the canopy above will be diverted through a platform drain to the connecting drain pipe below

6. Wet Column: Slices within the column reveal an inner textured surface that directs water to the drain below. The cut-out also provides visual and lighting interest for those awaiting the next train.

Credit: Anuj Kumbhar

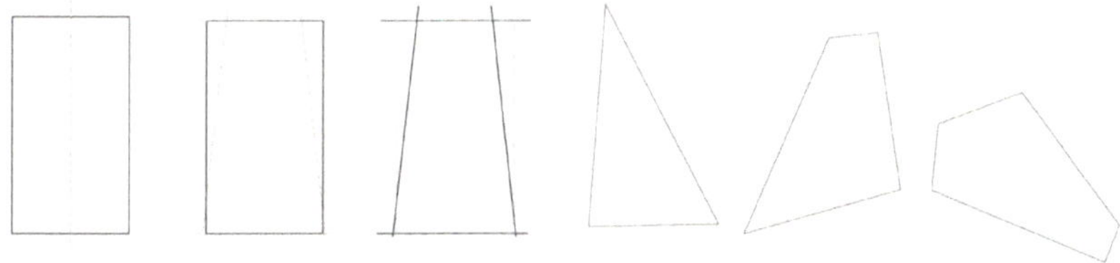

As its corner or edgy sape and orientation enables it to reduce the drag and the air ,oves thrigh the structure further minimising te notice levels and this form helps in avoiding paralal air movement.

• Redefine the bus terminal as destination rather than an interesting stop
• The problem looks at an idea that each bus terminal stand in isolation.
• Bus terminals are getting transformed from being monotonous structure to lively places of social gathering and entertainment.
• Bus terminal become an important landmark and welcoming gateway to commuters Visiting the city it adds to the image of the city for travellers community to and from the city.
• The idea is that there is free flowi in the movement of everyone without entering into others property zone.

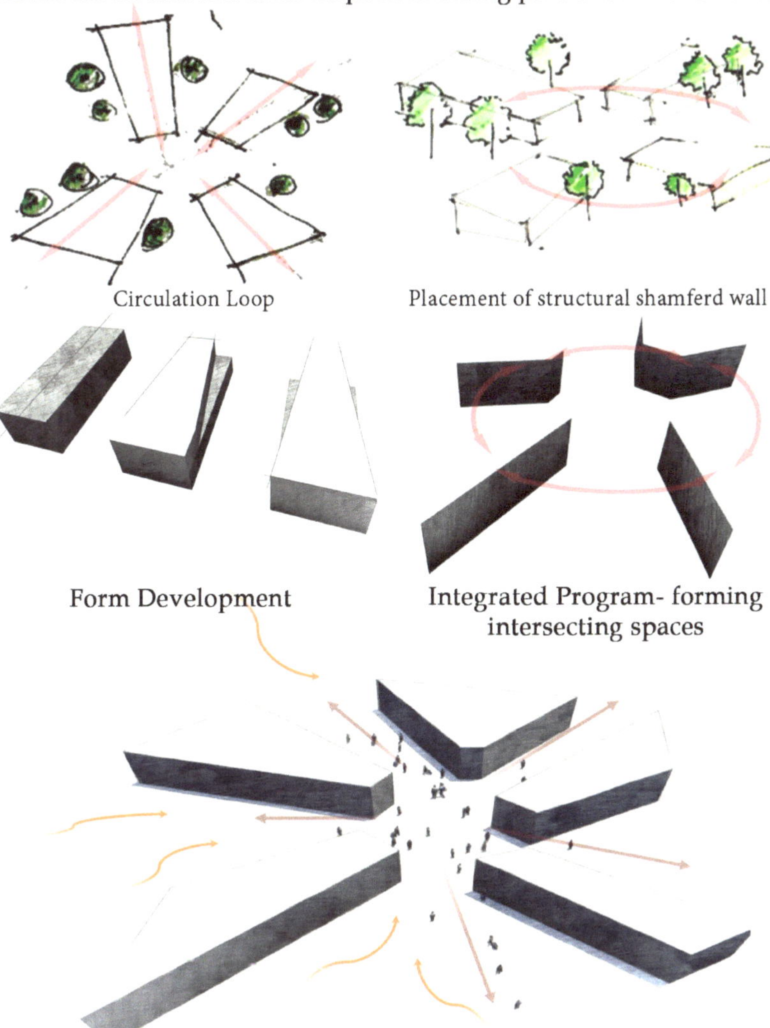

Circulation Loop

Placement of structural shamferd wall

Form Development

Integrated Program- forming intersecting spaces

Movement between the spaces

PLAN

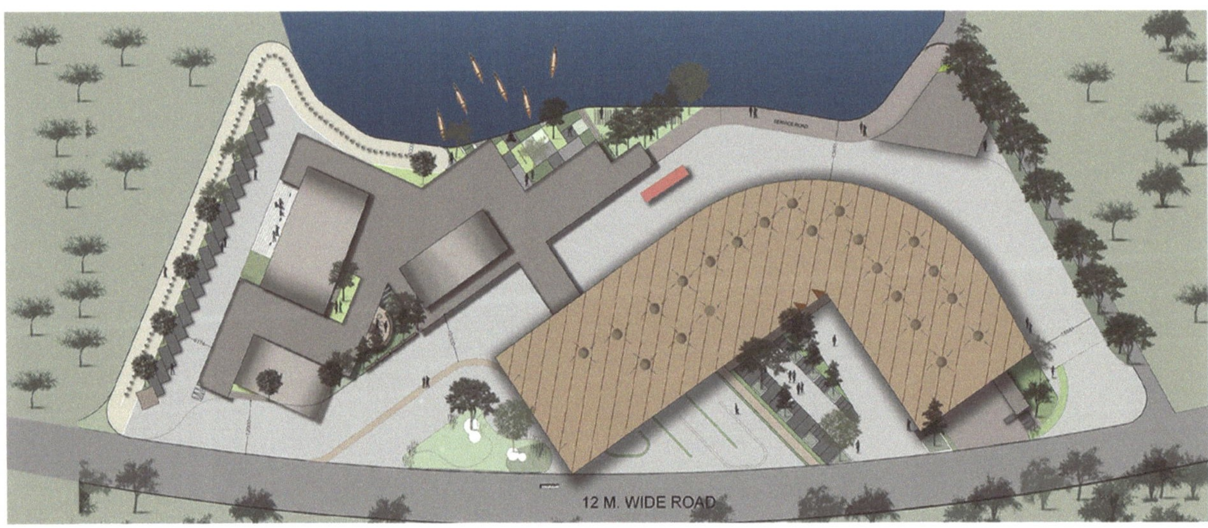

ROOF PLAN

SECTION

SECTION

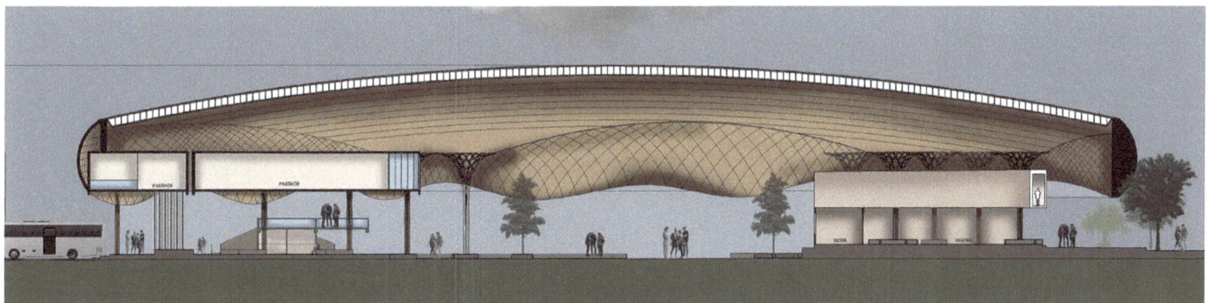

SECTION

EAST SIDE ELEVATION

INDOOR CONNECTION
WITH GREEN SPACE

CONCEPTUAL CONTEXT SKETCHES

Integrated bus terminal with sports complex
(Valvan lake)

The concept is derived from Biophilla. A hypothetical human tendency to interact or br closely associated with other forms of life in nature. A desire or tendency to communicate with nature. 2 cortyard genric plan connection with outdoors where to access fresh air, presence of water, green space and day light.

GROUND FLOOR

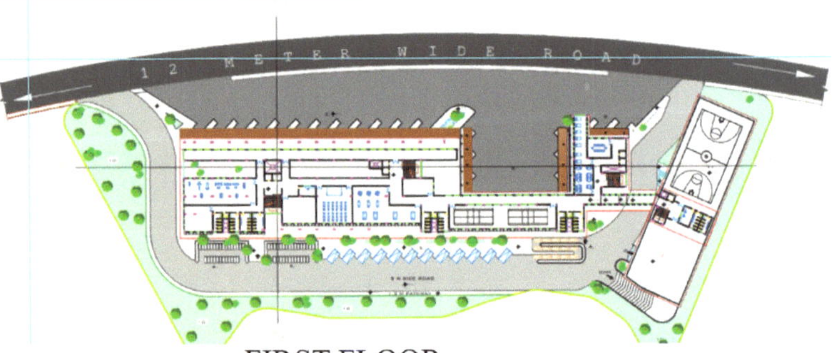

FIRST FLOOR

SECTION A-A'

WEST SIDE ELEVATION

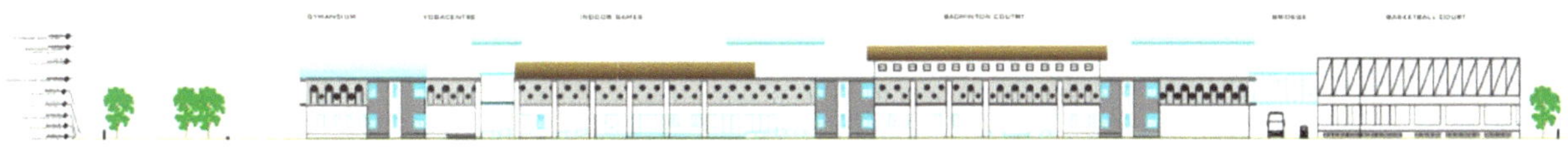

SOUTH SIDE ELEVATION

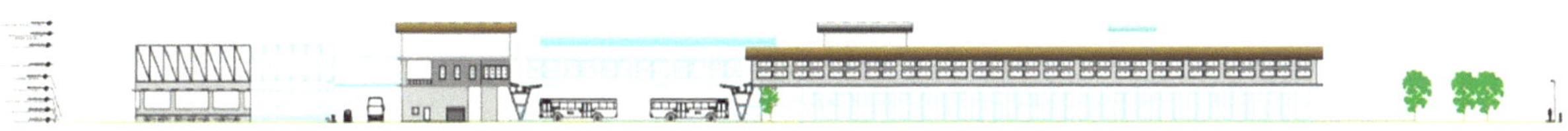

NORTH SIDE ELEVATION

Integrated bus terminal with sports complex
(Valvan lake)

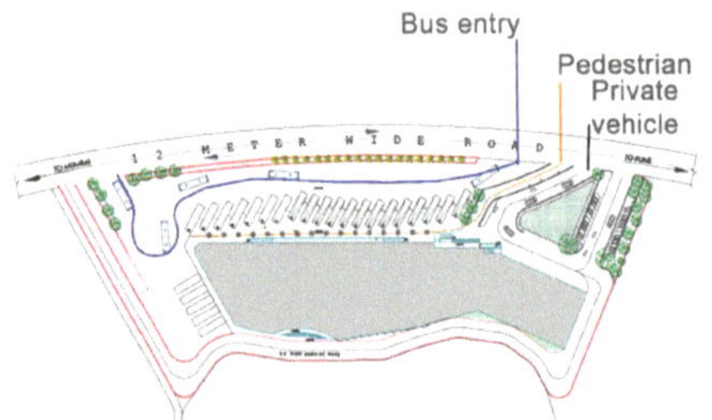

Circulation of bus terminal is major issue.

The circulation is segregated by providing 3 different entry for bus, pedestrian, and feeder service/private vehicle.

CONCEPT

TRIBUTE TO TRIBUTARY

The concept is derived from the history of indrayni river, it is the sacred river of lonavala which is a tributary of bhima river, which further merges to krishna river, the river flows from two cities dhenu and alandi, which is associated with the sacred names of two great saints viz Sant Tukaram and Sant Dyaneshwar. Due to industralization and other activities this river was polluted to preserve it valvan reserviour was constructed by tata industries.

Choosing the concept of flowing river and fuildity in planning to develop a neo-futuristic form which tributes the indrayani river and comoliment its surrounding.

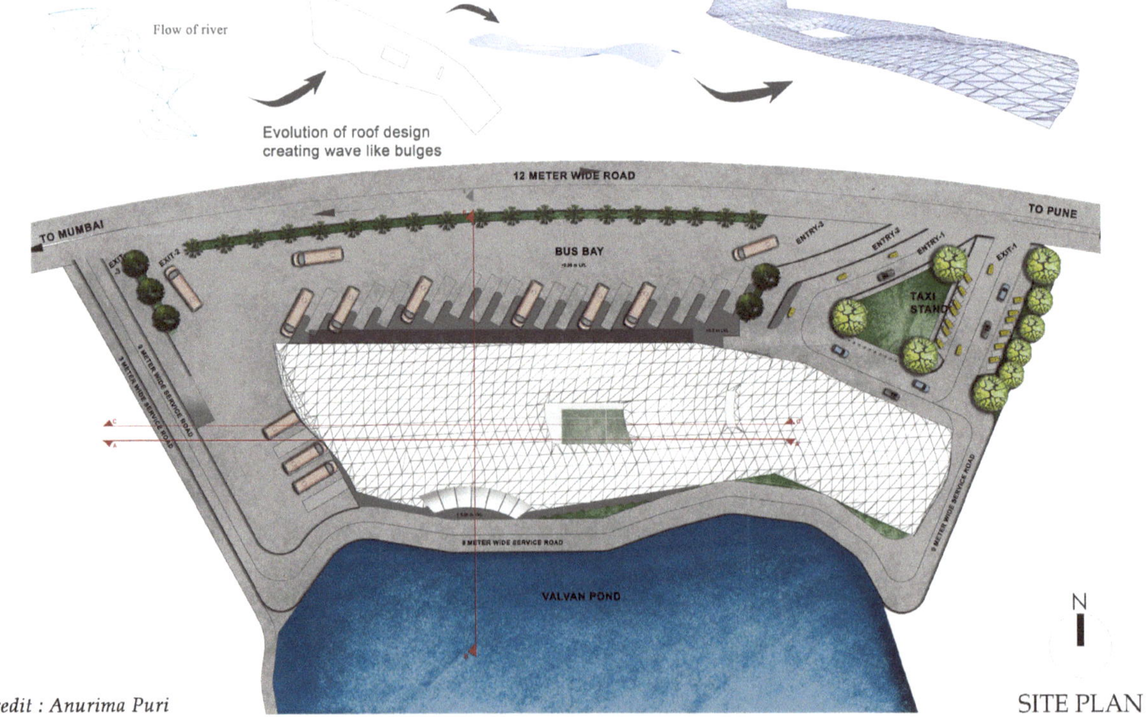

Credit : Anurima Puri

Section AA′

Section BB′

Basement Plan

GROUND FLOOR PLAN

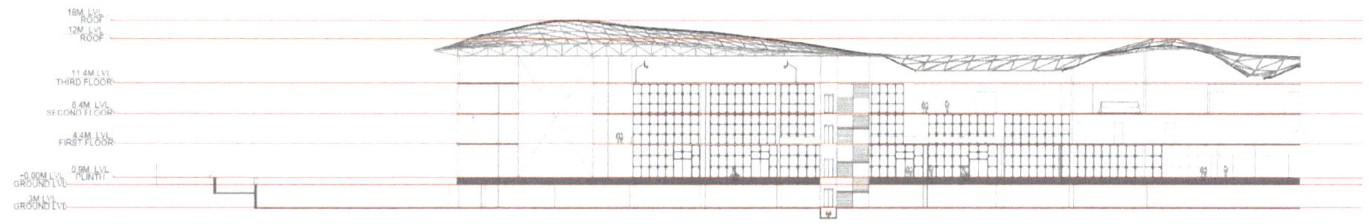

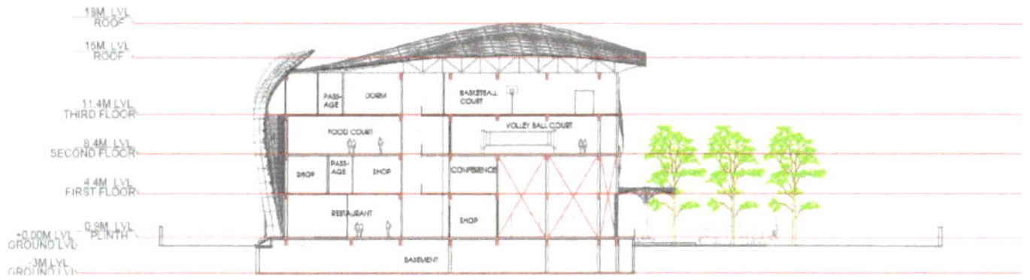

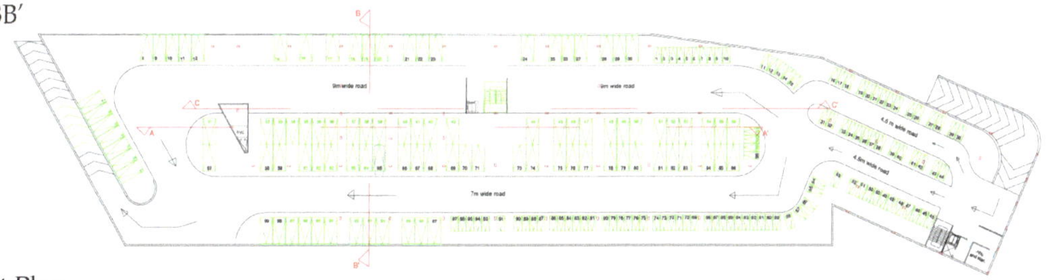

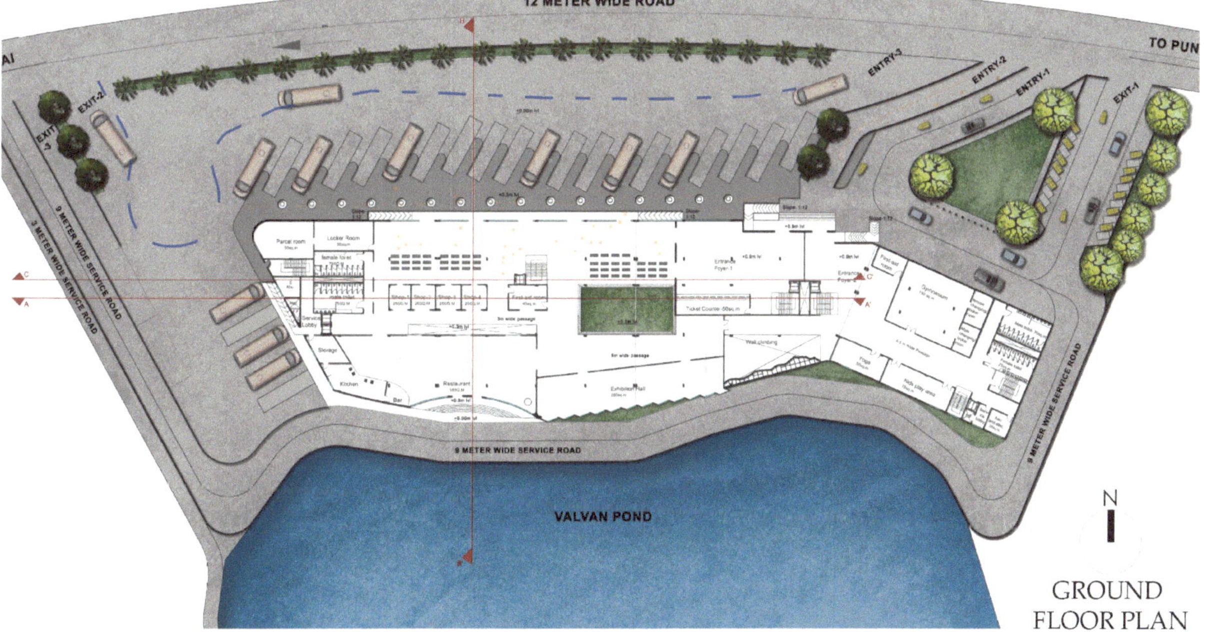

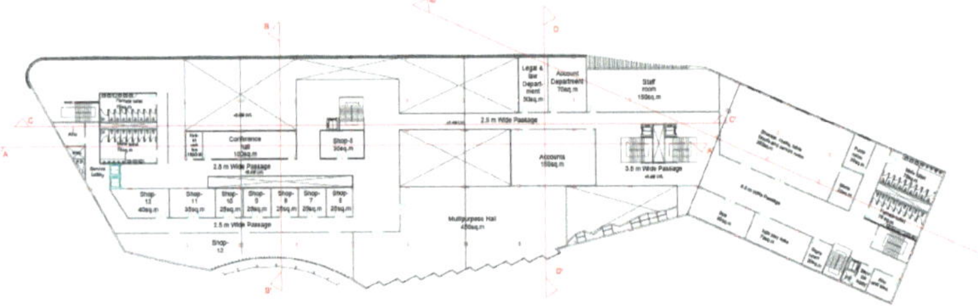

First Floor Plan

View 1

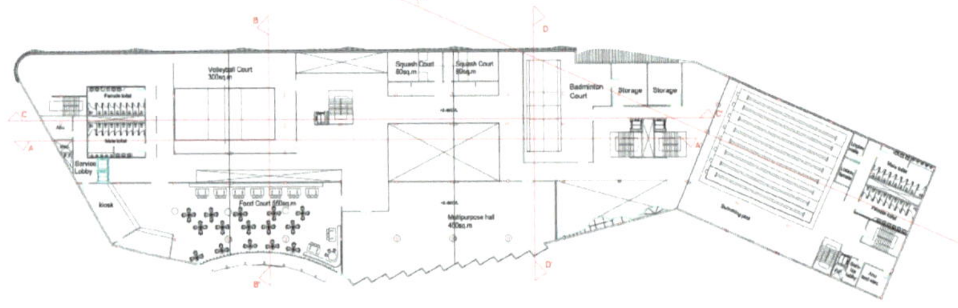

Second Floor Plan

Front Facade View

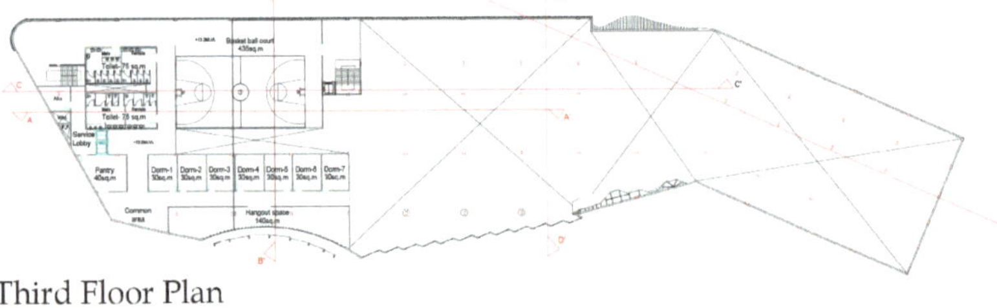

Third Floor Plan

View 2

View 3

PLANS
VIEWS

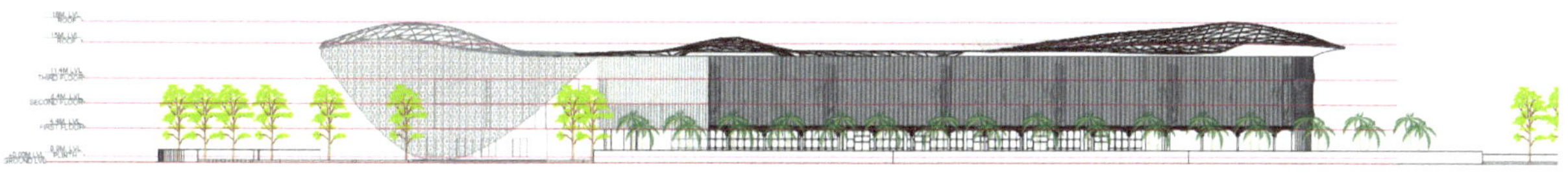

Front Elevation

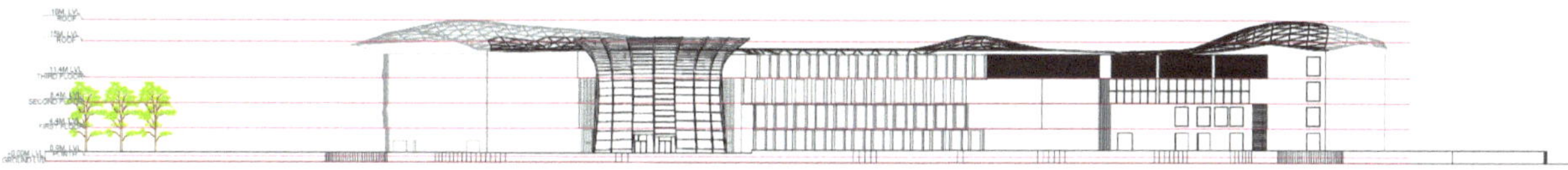

Back Elevation

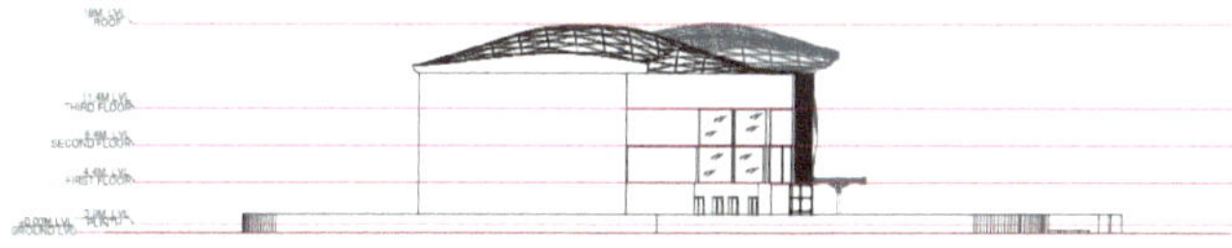

Left Side Elevation

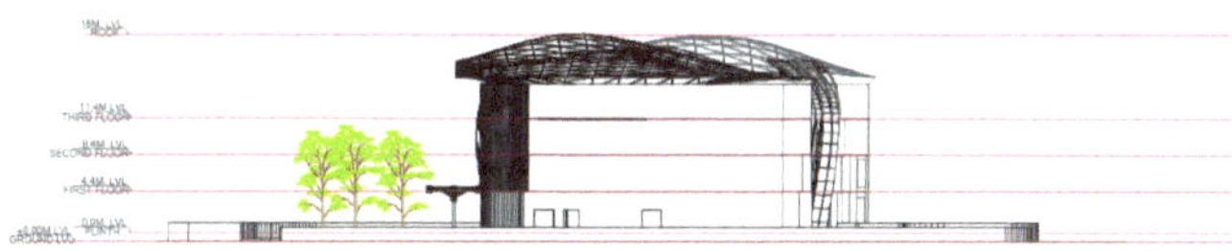

Right Side Elevation

View 4

View 5

ELEVATIONS
VIEWS

125

CONCEPT:

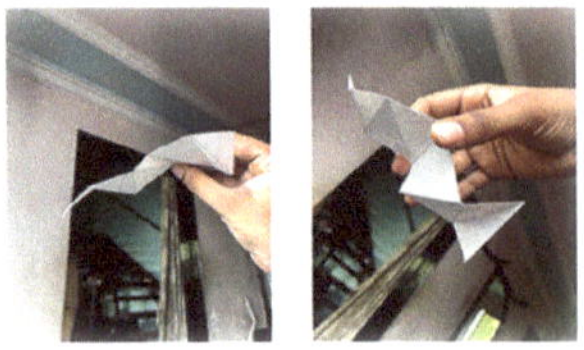

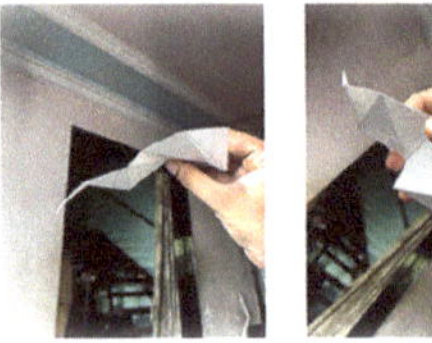

Integrated bus terminal with sports complex

(Valvan lake)

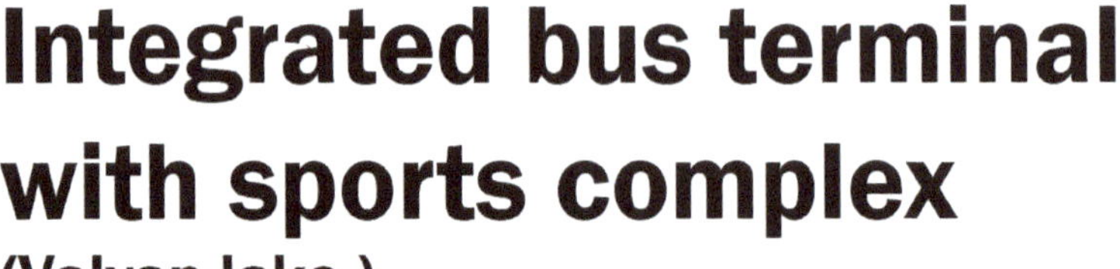

Origami is the art of paper folding. In the future, this building would become a architectural landmark, making it outstanding and iconic and it would stand out in the environment, which is why I chose the origami design.

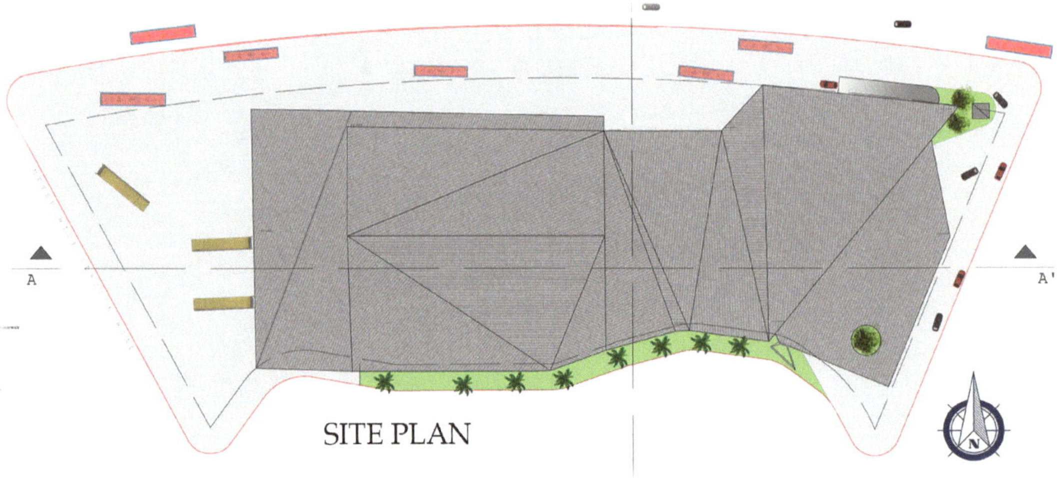

SITE PLAN

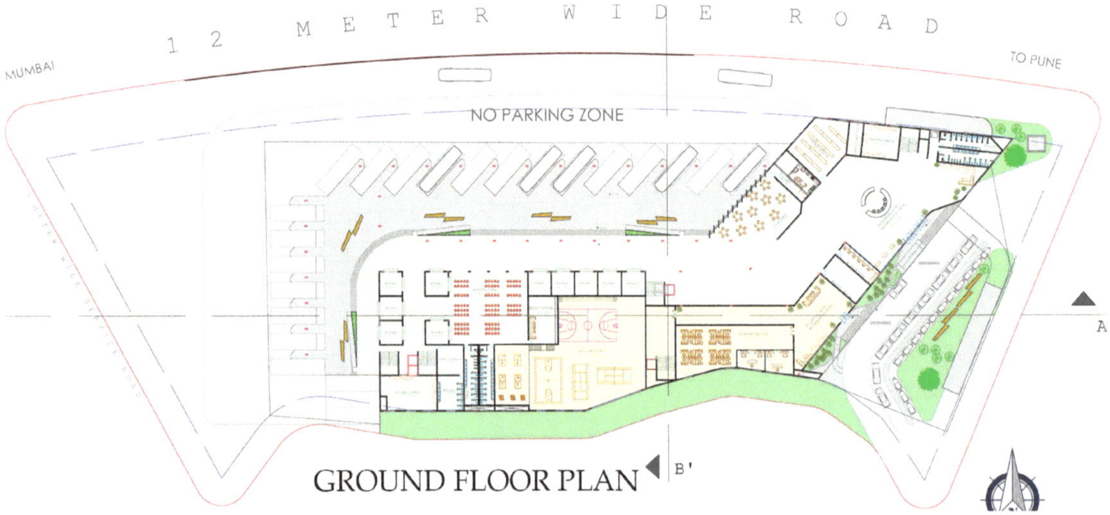

GROUND FLOOR PLAN

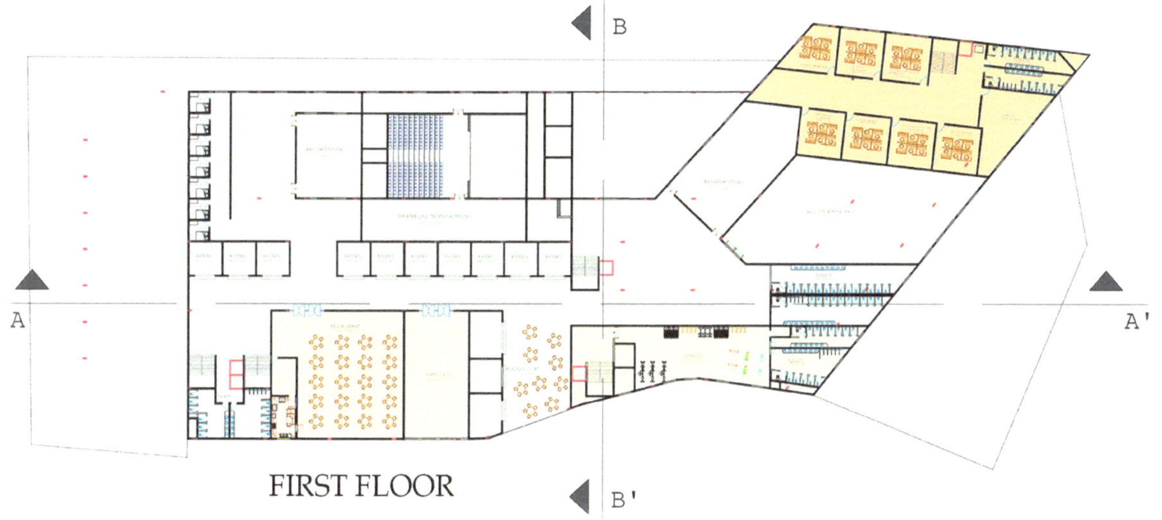

FIRST FLOOR

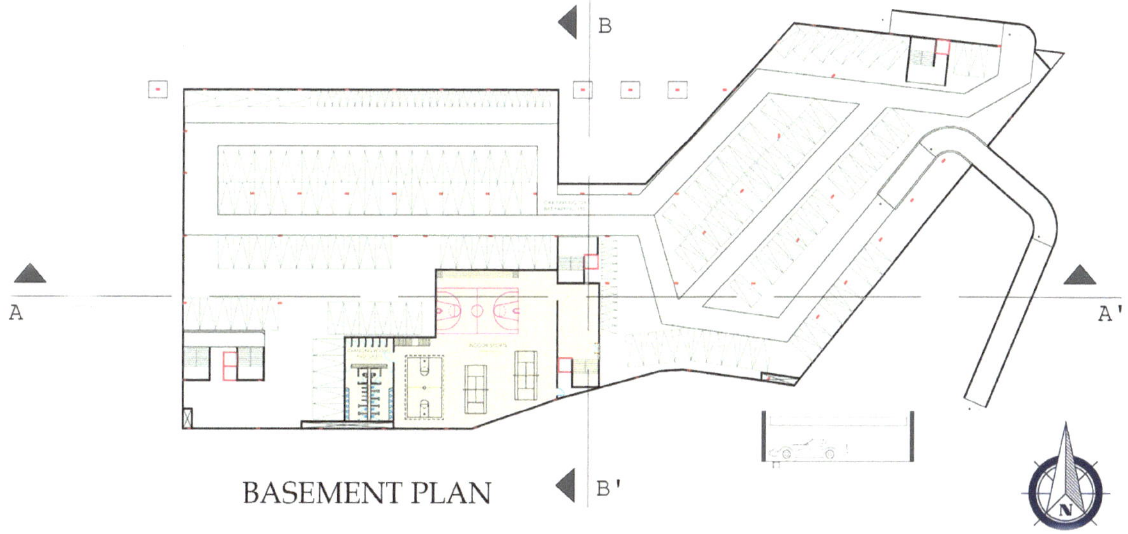

BASEMENT PLAN

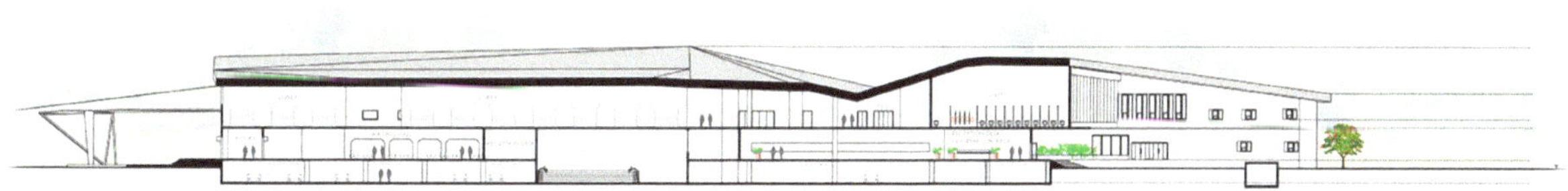

SECTION A-A′

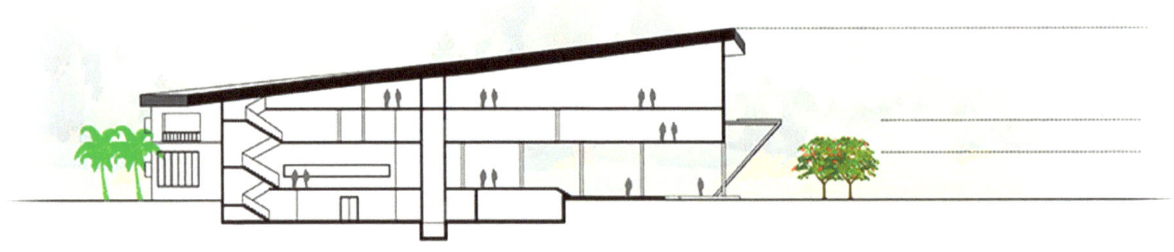

SECTION B-B'

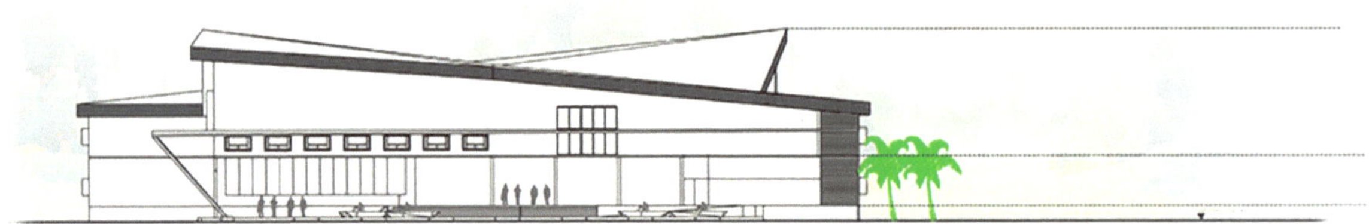

WEST ELEVATION

EAST ELEVATION

LIST OF IMAGES

Image 1: Loanvala dam

Image 2: Waterfall in Loanvala

Image 3: Tunnels and road network carved through the controus of the region with the settlement spread over the region.

Image 4: Bhor ghat, Khandala -2023

Image 5: Bhor ghat, Khandala -1895

Image 6 :This is the lonavala dam ,which is a watershed area where water from the surrounding mountains is channelled into the dam . Also the dam is build on the Intrayani river which furthure flows into the city and merges with the bhima river.

Image 7: Tata dam along the sided of Lonavala lake, 2023.

Image 8: Connecting higways and expressways to Lonavala, 2023.

Image 9: Khandala Ghat

Image 10: Ryewood Park

Image 11: Likely seen dense vegetation

Image 12: Central core of Lonavala

Image 13: Then Karli caves presently known as the famous Karla caves.

Image 14: Present day urban fabric showing recreational spaces- Rama Krishna hotel and Della adventures

Image 15: A network of major roads,railway line connecting to Lonavala.

Image 16: Complex road networks and interconnectivity at Center Point-an important node in lonavala

Image 17: Kumar resort junction at 8 AM in the morning.

Image 18: Shivaji Maharaj chowk at 8 AM in the morning.

Image 19: Lohgad Udyan at 8 AM in the morning.

Image 20: Movement patterns of tourist and local inhabitants.

Image 21: Famous tourist attractions.

Image 22 : Reminiscent of street shopping experienced during Friday market, 2023.

Image 23: A perspective of Friday market street depicting the essence of a vibrant shopping experience, 2023.

Image 24 : A perspective of Friday market at Purandar ground depicting the essence of a vibrant shopping experience, 2013.

Image 25: Jaichand chowk - A Prominent node.

Image 26: From Shivaji Maharaj chowk towards Bhangarwadi

Image 27: From Shivaji Maharaj chowk towards Gaonthan

Image 28: View of Khandala, a small hill station in the state of Maharashtra, 1860s.

Image 29 : A waterfall near Lonavala lake.

Image 30: Narayani Dham

Image 31: Waterfall near Lonavala lake

Image 32: Sterling hotel, Lonavala

Image 33: Meher villa resort, Lonavala

Image 34 :Maganlal and sons chikki shop -2023

Image 35 :Maganlal and sons chikki shop -

Image 36 :Bhaji Mandai

Image 37:Arial view of the built environment

Image 38 : Lane parallel to Shivaji Chowk showcasing naturally evolved mixed-use community

Image 39:Main bazaar road leading towards Mumbai-Pune highway

Image 40: Shivaji chowk

Image 41: Arial view of the Indrayani river

Image 42: Valvan lake

LIST OF FIGURES

REFERENCES

1. Survase, M. (2013). Assessment of socio-economic and business profile of Chikki trade promoters in Lonavala.
2. Government of India. (2011). Census of India. Retrieved from https://www.census2011.co.in/data/town/802810-lonavala-maharashtra.html#google_vignette
3. Lonavala Square. (2019). Catchment report.
4. Bombay Natural History Society. (2003). Datazone: Shivaji and adjoining areas, Lonavala IBA, India. Retrieved from https://datazone.birdlife.org/site/factsheet/ins--shivaji-and-adjoining-areas-lonavla-iba-india
5. Lynch, K. (1960). The image of the city. Cambridge, MA: MIT Press.
6. McHarg, I. L. (1969). Design with nature. New York: Wiley.
7. Orr, D. W. (2002). The nature of design: Ecology, culture, and human intention. Oxford: Oxford University Press.
8. Forman, R. T. T. (2014). Ecology of urbanization: Worldwide impacts. Cambridge: Cambridge University Press.
9. Mehrotra, R. (2008). "Choreographies of the modern: The city and the landscape." Journal of Landscape Architecture, 3(1), 20-29.
10. Hawken, P. (1993). The ecology of commerce: A declaration of sustainability. New York: HarperBusiness.
11. Raghunandan, C. R. (2016). Reimagining India's urban future: Visionary ideas for urban design and architecture. New Delhi: Urban Design Publications.
12. Lall, A. K. (2012). Green design: From theory to practice. New Delhi: TERI Press.
13. Singhal, S. (2017). Cities of the future: A green design agenda. Mumbai: Green Urban Publications.
14. Forman, R. T. T. (2014). Urban ecology: Science of cities. Cambridge: Cambridge University Press.
15. Satterthwaite, D. (1999). Sustainable cities: Urbanization and the environment in the developing world. London: Earthscan.
16. Sundaram, R. S. (2020). Public spaces in urban India: Design, policy, and practice. Chennai: Urban Planning Press.

Mentors and Editors
Prof. Suchetha Mathew Thazhathil
Prof. Viji Nair
Prof. Swechcha Gosavi

Contributors
Ameya Bhagat, Mangesh Dervankar, Aarya Deshpande, Rashmi
Deshpande, Savi Gosavi, Ritesh Igave, Snehal Jangam, Chetan
Joshi, Pranali Kadav, Anushree Kamble, Prayag Khandalkar,
Dharmika Khobrekar, Disha Kothari, Anuj Kumbhar, Gayatri
Mahajan, Prathamesh Mane, Neil Mankar, Aoudhut Mhatre,
Akshata Patil, Hrushikesh Patil, Anurima Puri, Rahul Rana,
Shreyas Salunke, Heena Sharma, Shyammurat Vishwakarma, Tejas
Vishwakarma, Sumedh Vispute, Manish Walekar, Veda Warge,
Dipak Palwe, Srushti Ingle, Ruchita Kadam.

Compilation team
Team head: Aarti Gawade
Members: Janhavi Nikam, Anuradha Patil, Shubham Patil,
Anushree Pednekar, Purva Salunke, Yash Singh, Dhiraj Thorat,
Rushabh Nalawade, Pranita Mhatre.